T0382921

Methods in Product Design

New Strategies in Reengineering

Engineering and Management Innovation

Series Editors

Hamid R. Parsaei and Ali K. Kamrani

RECENTLY PUBLISHED

Methods in Product Design: New Strategies in Reengineering
Ali K. Kamrani, Maryam Azimi, and Abdulrahman M. Al-Ahmari

Systems Engineering Tools and Methods
Ali K. Kamrani and Maryam Azimi

Optimization in Medicine and Biology
Gino J. Lim and Eva K. Lee

Facility Logistics: Approaches and Solutions to Next Generation Challenges
Maher Lahmar

Methods in Product Design

New Strategies in Reengineering

Edited by
Ali K. Kamrani
Maryam Azimi
Abdulrahman M. Al-Ahmari

CRC Press
Taylor & Francis Group
Boca Raton London New York

CRC Press is an imprint of the
Taylor & Francis Group, an **informa** business

To my aunt, Fakhrie

—**Maryam Azimi**

To our students

—**Ali K. Kamrani**
—**Abdulrahman Al-Ahmari**

Contents

Preface

The current marketplace is undergoing an accelerated pace of change that challenges corporations to apply new techniques to respond rapidly to this ever-developing environment. At the center of this change is a new generation of customers. As the industry adopts a consumer focus in its product development strategy, it must offer broader product ranges, shorter model lifetimes, and the required ability to process products in less time and costs. A consumer-focused product design must simultaneously meet the conflicting objectives of consumer and manufacturer. It is based on premises that (a) changing customer requirements dictate varied product features, (b) the structure of products and processes must be aligned with dynamic product features, and (c) manufacturing productivity requires managing conflicting objectives due to these structural alignments.

Organizations now fail or succeed based upon their ability to respond quickly to changing customer demands and to utilize new technological innovations. In such an environment, the advantage goes to the firm that can offer greater varieties of new products with higher performance and more overall appeal. In order to compete in this fast-paced global market, organizations need to produce products that can be easily configured to offer distinctive capabilities compared to the competition. Furthermore, organizations need to develop new methods and techniques to react rapidly to required changes and to shorten the product development cycle, which will enable them to gain more economic competitiveness. This edited book is a collection of methods and state-of-the-art technologies in new strategies for customer-focused product design and development.

Chapter 1 by Ayyamperumal et al. introduces the concepts of sustainable design, how to quantify and calculate environmental impact metrics, and the commercially available tools that help a design engineers to create sustainability designs. It also provides the knowledge needed to use sustainability models and tools to explore trade-offs between eco-friendliness and cost. As one of the most important applications of group technology in manufacturing, cellular manufacturing seeks to deliver high productivity and flexibility for manufacturing different product varieties. Cell formation of part families and machine cells is the critical element in designing an efficient cellular manufacturing system. In Chapter 2, Zhang provides an advanced survey of methods for CMS design. As an advanced tool, CAD has been used for design of complex systems. Chapter 3 by Kamrani presents an overview discussion on Computer-Aided Design (CAD) and feature representation methodologies. CAD and the supporting methods are used to facilitate integrated engineering design process. El-Tamimi et al. present a method for the parameter selection for CAD-VR data translation. To select the appropriate set of parameters, Design of Experiments (DOE) techniques are applied. Based on the statistical analysis of the selected parameters, a set of guidelines is developed for parameters selection during the conversion process. This is presented in Chapter 4. In Chapter 5, Nasr et al. propose a proposed framework of the integrated system for CAD and Computer-Aided Inspection (CAI). CAIP is based on the Automatic Features Extraction Module (AFEM), Computer-Aided Inspection Planning Module (CAIPM), and Coordinate Measuring Machine Module (CMMM). A case study is also presented to demonstrate the capability of the integrated system. Geometric modeling is used for design of complex shapes such as tumors. Chapter 6 by Azimi et al. presents results of an ongoing research in the development a three-dimensional (3D) model for tumor deformation predication during radiation treatment. MATLAB® software and rapid prototyping technology are used for modeling and validation of the predicated geometrical models. Product variety and its impact on manufacturing complexity is presented in Chapter 7. In this chapter, Kamrani provides an overview of different methods for measuring the degree of variety and complexity as proposed by other researchers. Chapter 8 continues the discussion on variety and manufacturing complexity by providing a new methodology. This chapter by Kamrani et al. presents a sample case study for analyzing manufacturing complexity due to increased product variety. A cost model is developed that captures the impact of increased inventory and the storage cost of subassemblies due to increased product variants. This is accomplished by generating a mixed model assembly sequence that aims to minimize the variation of subassembly inventories of the production span. Hassan et al. focus on the logistics issues, including the classification of the supply chain based on logistics networks. In Chapter 9, mathematical modeling of a dynamic SCN is developed, including a proposed solution. A case study on dynamic supply chains from a ready-mixed concrete (RMC) industry is presented. Chapters 10 and 11 are based on the maintenance planning. Maintenance is defined as the combination of activities and actions that are required to control and supervise a system to perform the intended functions. In Chapter 10, Al-Shayea proposes alternative methods for reducing maintenance and cost due to shut down and loss of productivity. In Chapter 11, Smadi develops a five-step methodology for preventive maintenance based on product nonconformances and quality. A case study is also presented in this chapter.

We would like to thank our authors and reviewers that participated in this project. We would also like to thank Auerbach for publishing this edited book, and express our appreciation for the opportunity given by our friend, the late Ray O'Conner, for enabling us to fulfill our vision in implementing this project.

Editor Bios

Ali K. Kamrani is an associate professor of industrial engineering at the University of Houston, where he is founding director of the Design and Free Form Fabrication Laboratory. He received his BS in electrical engineering in 1984, master's in electrical engineering in 1985, master's in computer science and engineering mathematics in 1987, and PhD in industrial engineering in 1991, all from the University of Louisville, Louisville, Kentucky. His research has been motivated by the fundamental application of systems engineering and its application in advanced design and development of complex systems.

Maryam Azimi is a software project manager at Lenovo Corporation. She received her PhD in industrial engineering from the University of Houston in 2011. Her research interests are systems engineering, data mining in health care, Lean Sigma Six, and project management.

Abdulrahman Al-Ahmari is a professor of industrial engineering at King Saud University, Saudi Arabia, where he serves as dean of the Advanced Manufacturing Institute. He received his PhD (manufacturing systems engineering) in 1998 from the University of Sheffield, UK. His research interests are in analysis and design of manufacturing systems, computer integrated manufacturing (CIM), optimization of manufacturing operations, applications of simulation optimization, FMS, DOE, and cellular manufacturing systems.

Contributors List

Chapter 1

Pratheep Ayyamperumal
Mechanical Engineer
Boston, Massachusetts

Ranjit Vinu
Boston, Massachusetts

Ibrahim Zeid
Department of Mechanical and
    Industrial Engineering
Northeastern University
Boston, Massachusetts

Sagar Kamarthi
Department of Mechanical and
    Industrial Engineering
Northeastern University
Boston, Massachusetts

Tucker J. Marion
Entrepreneurship and Innovation Group
D'Amore–McKim School of Business
Northeastern University
Boston, Massachusetts

Chapter 2

Yaowu Zhang
Corporate R&D engineer
Powell Industries, Inc.
Houston, Texas

Chapter 3

Ali K. Kamrani
Industrial Engineering
University of Houston, Houston, Texas
    Graduate Program Studies

Chapter 4

Mustufa H. Abidi
Industrial Engineering Department
College of Engineering
King Saud University,
Riyadh, Saudi Arabia

Abdulaziz M. El-Tamimi
Industrial Engineering Department
College of Engineering
King Saud University
Riyadh, Saudi Arabia

Emad S. Abouel Nasr
Industrial Engineering Department
College of Engineering
King Saud University
Riyadh, Saudi Arabia
and
Faculty of Engineering
Mechanical Engineering Department
Helwan University
Cairo, Egypt

Chapter 5

Emad S. Abouel Nasr
Industrial Engineering Department
College of Engineering
King Saud University
Riyadh, Saudi Arabia
and
Mechanical Engineering Department
Helwan University,
Faculty of Engineering
Helwan, Cairo, Egypt

Abdulrahman Al-Ahmari
Advanced Manufacturing Institute
Industrial Engineering Department
King Saud University
Riyadh, Saudi Arabia

Osama Abdulhameed
Industrial Engineering Department
(master's degree student)
King Saud University
Riyadh, Saudi Arabia

Chapter 6

Maryam Azimi
Software Project Manager
Lenovo Corporation

Ali K. Kamrani
Industrial Engineering Department

Emad Samir Abdelghany
Industrial Engineering Department
College of Engineering
King Saud University
Riyadh, Saudi Arabia
and
Faculty of Engineering

Mechanical Engineering Department
Helwan University
Helwan, Cairo, Egypt

Chapter 7

Ali K. Kamrani
Industrial Engineering Department
University of Houston, Houston, Texas

Chapter 8

Ali K. Kamrani
Industrial Engineering
University of Houston, Houston, Texas

Arun Adat
Supply Chain Strategy and
    Development Manager
Hewlett Packard
Houston, Texas

Maryam Azimi
Software Project Manager
Lenovo Corporation
Morrisville, North Carolina

Chapter 9

Mohammed Hussein Hassan
Industrial Engineering at the Faculty
    of Engineering
Helwan University
Cairo, Egypt

Haitham Abbas Ahmed Mahmoud
Industrial Engineering at the Faculty
    of Engineering
Helwan University
Cairo, Egypt

Chapter 10

Adel Al-Shayea
Industrial Engineering Department
College of Engineering
King Saud University
Riyadh, Saudi Arabia

Chapter 11

Hazem J. Smadi
Industrial Engineering
Jordan University of Science and
    Technology
Amman, Jordan

Chapter 1

Sustainable Design

Pratheep Ayyamperumal, Ranjit Vinu, Ibrahim Zeid,
Sagar Kamarthi, and Tucker J. Marion
Department of Mechanical and Industrial Engineering, Northeastern University

Contents

1.1 Introduction

New products are designed and developed to serve ever-changing and demanding human needs. Until recently, engineering design did not consider the environmental impact of products throughout their life cycle. As a result, though customer needs were served, undesirable environmental impacts were caused. Engineering design is now more environmentally conscious by incorporating the concept of sustainability in New Product Development (NPD). Incorporating sustainability leads designers to investigate the environmental impacts caused by the product, starting from its design through commercialization and disposal. Sustainable design (also known as eco or green design) is becoming a mainstream philosophy in engineering practice.

This chapter introduces the concepts of sustainable design, how to quantify and calculate environmental impact metrics, and the commercially available tools that help a design engineer to create sustainability designs. The chapter also covers the product life cycle and life cycle assessment (LCA) to help explain sustainable design. The chapter also provides the readers with the knowledge needed to use sustainability models and tools to explore trade-offs between eco-friendliness and cost.

1.2 Engineering Design

Engineering design is always directly impacted by societal needs. The basic premise of engineering design is to create products of high quality at reduced cost. However, design paradigms change with time, depending on market forces. Some of the well-known paradigms are DFA (Design for Assembly), DFM (Design for Manufacturing), DFX (Design For Anything), and CE (Concurrent Engineering). DFA attempts to streamline product design to be easy to assemble. DFM focuses on the manufacturability of the design. CE attempts to shorten the design cycle by considering all aspects of design at the same time, by avoiding sequential design. For example, in adopting CE, a design team includes all types of engineers: design, materials, sales, marketing, and manufacturing. These engineers brainstorm and critique a design from all points of views, thus finalizing the design quicker and better and avoiding any pitfalls down later in the product cycle.

Sustainable design (also known as eco or green design) is the latest paradigm shift in design focus. This focus is the result of organized efforts on the part of governments and professional organizations to save the environment and the earth's natural resources, minimize energy consumption, as well as reduce the amount of waste, as landfills are rapidly reaching their full capacities (McDonough and Braungart 2002).

Sustainable design goes beyond ensuring that a design meets the functional requirements of a product. It looks at the full life cycle of the product, from "birth" to "death," also known as from "cradle to grave" (McDonough and Braungart 2002). Designers must answer the questions of how much energy a product consumes during its production and its useful life use; and what to do with the product when it reaches its end of life (EOL). EOL planning explores how to dispose of the product: recycle, remanufacture, or discard into landfills (Graedel and Allenby 1995; Ginley and Cahen 2011).

In order to gain a good understanding of sustainable design and its tools, we begin by covering sustainability and LCA in detail. We also quantify sustainability and provide the reader with tools to help assess design alternatives from a sustainability viewpoint.

1.3 Depletion of Natural Resources

Products consume natural resources in different shapes and forms. The only time a product does not consume natural resources is when it is in the design phase. Once a product design is finalized and the product is to be manufactured, the product begins to consume natural resources in the form of raw materials (Fu et al. 2000). Also, energy is consumed during the manufacturing processes. The product may also consume energy during its use. Moreover, the product consumes natural resources at its EOL for disposal (Manzini and Vezzoli 2002; Manzini and Jegou 2003).

The above steps cause the depletion of natural resources, and hence the product development process is the foundation for investigations and studies to ensure the durability and sustainability of the natural resources that products consume. The quest here is to identify how the material and energy used during product life cycle could be replenished so that the product design becomes sustainable (Field and Ehrenfeld 1995; Marten 2001).

Upon completing its designed lifetime, the product is scrapped. If the scrap could be reprocessed to regenerate materials needed to manufacture other new product, the raw materials start to sustain without depleting their availability (Cagno et al. 2000; Lambert and Gupta 2005). Further, if renewable energy sources were substituted in place of nonrenewable fossil fuels, depletion of energy sources can be avoided or minimized, eventually paving the way for sustainability of the energy resources.

Thus, apart from the product development process, the service and death of the product become significant for achieving sustainability. The qualitative and quantitative study of a product from its birth to incineration is called the Life Cycle Assessment (LCA) of the product. As the product on its death/incineration may help preserve sustainability of the resources, is clear that the life cycle of the product determines the sustainability of the raw materials and involved natural resources (Rees 1992). Therefore, LCA becomes an essential and an indispensable framework to quantify and facilitate sustainability.

1.4 Need for Sustainability

Sustainability means the ability to endure and survive without depletion. "The balance of nature" (Ecological Balance) is a theory that says that ecological systems are usually in a stable equilibrium ("homeostasis"). This implies that any change in a particular parameter governing the balance causes a corresponding/appropriate change in other parameters that govern the balance and thereby the original balance and equilibrium of nature is maintained. The supposed reactive changes are almost always undesirable for a better human living environment and are often unsuitable for a stable ecology (Fussler and James 1996).

As production increases, there are good chances for rapid depletion of the earth's resources as raw materials are consumed (Schmidt-Bleek 1995). When the raw materials are processed and used for manufacturing, not only is energy consumed, depleting the energy sources, but also considerable harmful wastes are generated that pollute the air, water, and soil. The product's life and its disposal also lead to environmental impacts in the form of energy consumption and pollution (Campbell et al. 2006).

Thus, the stages of the product life cycle such as material consumption, manufacturing, operation, and disposal are susceptible to disrupting the balance of nature. These stages provide good reasons for nature to alter itself to maintain its balance and therefore have a high potential to lead to undesirable ecological states. If the undesirable changes should be overcome, materials and energy should be consumed at *sustainable yield* (Simon and Dowie 1993). Sustainable yield is when the consumption is equal to or less than the regeneration of the resources, resulting in sustainability (United States 1995; Vigon et al. 1992).

To achieve sustainable yield, product life cycle assessment is critical in the way that material usage, energy usage, as well as impacts of manufacturing processes, product operation, and incineration on the environment can be studied and quantified to achieve sustainability. The steps to sustainability are thus crucial because they facilitate averting possible negative and undesirable impacts so that a desirable stable environment and ecological state is maintained.

1.5 Product Life Cycle

Once the material for the product is chosen and the manufacturing processes are defined, the product is ready for production. The tangible resources needed at this point are the materials, which are taken from the environment. However, the material is not available in a form that is ready to manufacture. It naturally comes in raw form and needs processing to make it suitable for manufacturing. This phase is the Pre-Production phase of the product life cycle.

After the material is processed to suit manufacturing, it enters the designed manufacturing processes and now the product life cycle is at its Manufacturing stage. This stage also covers assembly of components until the final product is ready to use (Field and Ehrenfeld 1995).

The ready-to-use manufactured product has to be moved from the shop floor to the market and subsequently to its place of operation. Logistics is a separate study in supply chain management. This stage of the life cycle can be termed transportation/distribution.

The service the product offers to the customer is put in the Operation/Use phase of the product life cycle. Upon serving its intended lifetime, the customer discards the product, which is then taken to recycle the possible materials and dispose of the remaining useless components. This phase is the disposal phase of the product life cycle. During the disposal phase, parts of the product are processed back (post-process) to the raw material state (Simon et al. 1992). This post-processed material is used again to make new products. The other parts are returned back to the environment by incineration in the form of energy and waste (Marten 2001). Thus, the product originates from material and dies reducing to material. This cycle is called the PLC (*product life cycle*) (McLean and Lave 1998, Solow 1992).

1.6 Impacts

In each of the stages of a PLC, the material is the determining controlling factor upon which the product's various operations are based. During each operation, energy is used. The energy source may either be renewable like wind power, hydropower, or solar power, or nonrenewable like fossil fuels and nuclear fuels. Thus, PLC accounts for depletion of natural resources in the form of nonrenewable fuel sources.

While tangible resource consumption is an obvious impact on the environment, there are less obvious impacts on the environment that take place during the product life cycle. During all the stages of the product life cycle, processing of materials results in hazardous wastes in solid, liquid, and/or gaseous forms. Solid wastes can be found in the form of partially burnt and nonusable raw materials and energy-less charcoal. Dumping solid waste at landfills spoils the land and hence impacts the ecology of these landfills. Liquid waste may take the form of coolants or chemicals used to purify material during manufacturing. These liquids spoil water bodies, groundwater, and arability of land. The main source of gaseous pollutants is the burning of fuel during energy consumption at each stage. The gas pollutants cause air pollution that causes air acidification, which in turn results in acid rain. This is harmful not only to humans but also to the facing flora and fauna (Kinross 2002).

Thus, it is evident that to address the solution needed to minimize environment impact, knowledge of the product life cycle is crucial so that the cause and effect of environmental impacts can be traced and studied. Then the necessary steps to minimize them shall be taken and implemented leading to sustainability of materials, resources, and the environment.

The impacts thus far discussed can be generalized and classified into five types: carbon footprints, energy consumption, air acidification, water eutrophication, and water footprints. In order to relate the PLC to the product design phase, these impacts must be quantified and provided as design tools for designers to use to evaluate and assess the sustainability of their designs. We refer to the five types of impacts as impact metrics. A software tool that implements these impact metrics would be valuable for designers to use to evaluate the "what if" scenarios to study the impact of their designs on the environment (Murayama et al. 1999).

1.7 Impact Metrics

1.7.1 Carbon Footprints

The measure of all greenhouse gases that are produced during the product life cycle is defined as the carbon footprint. It is measured in the units of kilograms or pounds or metric tons of carbon dioxide (CO_2e) equivalent. The chief greenhouse gas is carbon dioxide. Greenhouse gases also constitute water vapor, methane, ozone, and nitrous oxide. Since CO_2 is the chief greenhouse gas, the impact of all other greenhouse gases is generalized and represented in terms of CO_2 equivalents. Any positive value for the carbon footprint is responsible for global warming that leads to extinction of many species due to their inability to adapt to the rise in temperature. The corresponding changes in the ecology and ozone layer depletion are highly undesirable since they cause environmental instability (Odum 2007).

As an example, the emission of an automotive can be calculated using the following equation (http://sustainability.rice.edu/Content.aspx?id=2397):

$$\left(\frac{MilesDriven}{AutoFuelEfficiency} \times 19.36 \right) \div 2204.6 = CO_2 Emissions(tons) \tag{1.1}$$

As per the US Environmental Protection Agency (EPA), an automotive vehicle that consumes 11.76 gallons of gas per week accounts to 12,500 pounds of CO_2 per year. The global average on greenhouse gas emissions from household wastes such as aluminum and steel cans, plastic, glass, and paper is estimated to 1027 pounds of CO_2 per year. For more detailed information, visit www.epa.gov.

1.7.2 Energy Consumption

Energy consumption is the driving force in the PLC. Energy is consumed at every stage of product manufacturing, transportation, product use, and so forth. By energy consumption, we refer to all nonrenewable natural resources consumed and depleted from the environment. The unit of measurement of energy consumption is *megajoules*. Any positive value for energy consumption means that there is a corresponding depletion of nonrenewable energy sources.

As per US Energy Information Administration (EIA), the United States consumes 20% of energy (includes oil, gas, nuclear, renewable, thermal, etc.) in the whole world, which is estimated at 98.6 quadrillion BTU (British thermal unit); equivalent to 1.04×10^{14} MJ approximately.

1.7.3 Air Acidification

The emissions such as carbon dioxide (CO_2), sulfur dioxide (SO_2), and various oxides of nitrogen (NO_x) that are produced when fossil fuel is burned react with the water vapor available in the atmosphere and form respective acids such as carbonic acid (H_2CO_3), nitric acid (HNO_3), and sulfuric acid (H_2SO_4). This causes air acidification and acid rain. Acidification restrains the growth of trees and contaminates groundwater. The unit of measuring air acidification is kilograms of sulfur dioxide (SO_2e) equivalent.

Air acidification is commonly caused by automotive emissions (SO_2 and NO_x), power plants, and manufacturing industries (SO_2, NO_x, and non-methane volatile organic compounds (NMVOCs)). Even domestic consumption such as cooking gas and air-conditioning systems cause air acidification. As per EPA, street traffic alone is responsible for approximately 50% of NO_x and NMVOC emissions. In 1994, United Nations Economic Commissions for Europe launched the Oslo Protocol to abate sulfur emissions. The protocol focuses on reducing sulfur content in fuel leading to SO_2 emissions and finding alternative oil derivatives. It has been proved efficient on a large scale so far. Visit http://www.unece.org/env/lrtap/fsulf_h1.html for more details.

1.7.4 Water Eutrophication

Water eutrophication is caused by dumping into water bodies the wastes generated from material extraction processes, manufacturing processes, and agricultural fertilizer wastes containing phosphates and nitrates. In certain conditions, industrial waste may be carriers of eutrophication agents. Emission by cars and trucks are also significantly responsible for water eutrophication. All processing industries, thermal power plants, pesticide industries, metal plating industries, and manufacturing industries produce toxic wastes and wastewater that contain phosphates (PO_4), nitrogen (N), calcium carbonate ($CaCO_3$), suspended solids, chlorides, and so forth. These wastes, when drained into water resources, contribute to water eutrophication. These wastes can cause serious ill effects to the water ecosystem and can cause environmental damage. Most power plants and industries have wastewater treatment units prior to disposal. The impacts occurring to water bodies are commonly measured in terms of kilograms of phosphates (PO_4e) equivalent.

Water eutrophication can be computed using the following equation:

$$TNI = \sum W_j^{TNI_i} \tag{1.2}$$

where

$$W_j = r_{ij}^2 + \sum r_{ij}^2 \tag{1.3}$$

Water eutrophication is generally computed on the basis of the Total Nutrient status Index (TNI) of water resources such as lakes and rivers. TNI is the sum of indexes of all nutrient parameters j, W_j is the proportion of j parameter in TNI, and r_{ij} is the relation of chlorophyll to other parameters. Visit the National Center for Biotechnology Information Journal at http://www.ncbi.nlm.nih.gov/pmc/articles/PMC2266883/.

1.7.5 Water Footprints

Each stage of the PLC consumes water in considerable volumes. This results in depletion of freshwater as a natural resource, causing shortage of freshwater. Since humanity relies on freshwater, water consumption is an impact metric for sustainability and measured in cubic meter (m^3) per year. The water footprint is an explicit indicator, showing volumes of water consumed.

As per the water footprint network, the global average of water footprint for generating nonrenewable energy resources such as natural gas, coal, crude oil, and uranium are 0.11 m^3, 0.16 m^3, 1.11 m^3, and 0.09 m^3, respectively, and for renewable resources such as wind energy, thermal energy, and hydropower are 0.00 m^3, 0.27 m^3, and 22 m^3, respectively.

1.8 Life Cycle Assessment

Life Cycle Assessment (LCA) is a method for measuring the environmental impacts associated with all stages of the PLC. The quantification of the impact metrics via the LCA helps us evaluate the environmental impact of a product. Thus, the LCA stages (steps) are

- Identify all interactions between a given activity in the PLC and the environment.
- Quantify each of the interactions in terms of the negative impacts on the environment using the standard impact metrics.
- Evaluate the total load on the environment using the quantification that covers all interactions of the PLC with the environment.

Finding the interaction of each stage of the PLC with the environment can be represented as shown in Table 1.1.

1.9 Design for Sustainability

Having assessed the PLC through the LCA, we now turn our attention to linking design and sustainability. We know that the more the CO_2, SO_2, NO_x emissions (gaseous oxides of carbon, sulfur, and nitrogen), the higher the carbon footprint and air acidification. The more by-products go to the water bodies, the higher the water eutrophication. The more the water is consumed, the higher the water footprint. Thus, in general, the higher the undesirable impacts, the higher the impact metrics are. Therefore, in order to achieve sustainability and the desired eco-design, all the impact metrics have to be minimized.

Table 1.1 LCA Stages and Interactions

Step	LCA Stage	LCA—Interaction
1	Material consumption	Material taken from the environment.
2	Material processing	Consumes energy and gives out pollutants and effluents.
3	Transportation	Material taken from environment. For packaging and fuel consumption for logistics.
4	Usage	Fuel may be consumed (natural resource consumed as energy) as per the product use.
5	Recycle/incineration	Recycle may have a positive impact (recycle) and a negative impact (incinerate) metric.

Since the design of the product influences the entire PLC, the necessary and correct decisions have to be made through the stages of the design process so as to have a lower impact metric at each stage of the PLC. Thus, design for sustainability aims at minimizing the product's adverse effect on the environment, resulting in eco-design or green-design.

1.10 Design Considerations for Sustainability

This section provides some guidelines for designers on how to consider sustainability while creating their designs. These guidelines have been developed along the lines of what current CAD software provides.

1.10.1 Efficient Energy Consumption

Designers need to optimize energy consumption at every phase of the PLC. They also need to build products that are efficient in using the available energy. For example, electric/hybrid cars are more efficient and consume less energy that their traditional counterparts that run on gasoline only. They are also less pollutant to air.

1.10.2 Renewable Energy Sources

The main difference between renewable energy and nonrenewable energy is the regeneration time. Renewable energy sources like wind current and water potential are perennial, whereas nonrenewable energy sources like fossil fuel takes millions of years to regenerate. Therefore, it is important that a new product developed should be designed to use renewable sources such as wind energy, thermal energy, hydropower, or solar power. Making such choices can result in conservation of fossil fuels and minimize the impact on the environment, for an essential sustainable future. Renewable energy resources impact the environment in two positive ways. It conserves the use of natural resources (fossil fuel reserve) and is less harmful to the environment.

1.10.3 Appropriate Materials

Material selection is most critical in a PLC. All product manufacturing processes are decided based on the material. The material needs to be procurable with minimal use of natural resources, safe to the environment and recyclable. Reusing material (recycled) in product development is critical to sustainability. For example, automotive manufacturers must consider lightweight material in order to get better mileage, thus reducing the estimated impact during the operation phase of the automobile. Pipe material selection offers another example. PVC pipes are commonly used in construction and cause severe impact to the environment;

they can be replaced by biodegradable plastic such as PLA (Polylactic Acid—http://en.wikipedia.org/wiki/Polylactic_acid).

1.10.4 Efficient Manufacturing Processes

The selection of manufacturing processes and machines should be based on achieving higher energy efficiency. This would require the designer to be knowledgeable about materials and processes to be able to select the better material. For example, rapid prototyping would be a better choice over casting if it can produce the same part.

1.10.5 Quality Manufacturing

Quality is an integral component in product development not only in terms of cost but also in terms of sustainability. Better quality helps reduce scrap and waste that save material and hence minimizes impact. Focus must be on implementing Six Sigma and other stringent quality check techniques to reduce rework and scrap.

1.10.6 Quality Product and Longer Product Life

Products must be durable and long lasting. Better product quality and longer product life reduce the environmental impact associated with servicing and repairing the product. For example, devices that are less likely to be damaged due to a fall have far less environmental impact than those that would need repair. Repair means consuming energy and in turn means impacting the environment.

1.10.7 Reuse and Recycle

All products manufactured should be designed for recycle and reuse at their disposal after the designed service (Zussman et al. 1994). Manufacturers should consider building products that recycle to the full maximum extent at the end of their useful life. For example, mining ore to make steel and mining bauxite to make aluminum are expensive operations to the environment both in terms of depleting natural resources and pollution. Thus, recycling of related products alleviate these two effects. For example, IBM has initiated a silicon wafer and chip recycling program. The silicon waste from several manufacturing industries is collected and recycled and reused in building newer silicon wafers. Although these wafers cannot be used in production, they can be of use in testing and experimentation. Visit http://in.reuters.com/article/2007/10/30/idINIndia-30240320071030 for more details.

1.10.8 Efficient Transportation

Transportation currently runs mainly on nonrenewable resources. This phase can be resorted to alternate sources that run on renewable energy or efficient energy (electric source). Most automotive manufacturers prefer assembling the entire vehicle and transporting it to the distribution location. Manufacturers may rather consider transporting top-level parts to various locations and assembling them close to large distribution centers.

1.10.9 Material Management

The impact to the environment can be reduced to a great extent by managing material use during the product design phase (Ginley and Cahen 2011). Girders often have an 'I' cross section to a rectangular cross section for the required moment to be achieved with less material. Bricks are integral materials used in construction. Modern construction management techniques have led to the development of hollow bricks, which save raw material and provide the required strength. Another example of effective material utilization is designing plastic chairs with holes. This helps in material usage and reduces manufacturing cost. Carbon nanotubes are more commonly used today to improve the strength of lightweight materials.

1.11 Sustainability Assessment

1.11.1 The Material Effect

The final design of a product dictates the rest of life cycle from manufacturing to use to retirement. More specifically, the material choice of the product decides all its downstream phases. Material selection determines the pre-production processes, the manufacturing processes, and the EOL processing. The time to evaluate the impact of the product material on the environment is during its design phase (Warhurst 2002). Selecting different materials and comparing their environmental impact is the basic premise of sustainable design (Rosy et al. 1993).

1.11.2 Assessment of Sustainability

Once a product material is selected, the LCA helps us quantify the impact metrics. Section 6 shows the metrics and how we can calculate them. The LCA can be summarized as a sequence of steps:

1. Define product geometry and shape.
2. Choose the appropriate material.
3. Select the manufacturing processes.
4. Select the place of manufacture.
5. Estimate the logistics overhead to the environment.
6. Estimate the product lifetime, and service time impacts.
7. Estimate the incineration, and recycle impacts.
8. Compute the cumulative impact, and document the results.

This sequence provides us with the "nodes" of the LCA "tree" where the optimization of environmental impacts can be performed. Logically, sustainability assessment deals with identifying the environmentally optimal value from the different values available at each of these nodes of the LCA so that the total measure of impact of the PLC is as low as possible, thereby leading to a more green and eco-design (Pennington et al. 2000).

During sustainability assessment, the LCA is evaluated for a selected material. The impact metrics for this first run is taken as the baseline (reference) of the sustainability analysis. Subsequent runs may be used to investigate other types of materials. The impact metrics of all the runs are compared, and the material with the least impact metrics is the winning design, resulting in a green design and hence a sustainable one.

1.11.3 Sustainability Software

If sustainability and the LCA are to become a useful design tool for designers to use, the impact metrics must be quantified, and this has been done; this makes them candidates that can be embedded in CAD software. Designers can then follow the concepts and steps covered in Section 10 to carry out their sustainable designs. There are a few software tools available to perform sustainability assessment such as Dassault Systems SolidWorks' Sustainability Xpress, PTC's Windchill Product Analytics, and PE International's GaBi LCA Software Platform. The steps that designers may follow to create sustainable design using a CAD system are as follows:

1. Create the part CAD model.
2. Assign material to the model.
3. Select a manufacturing process. CAD systems aid in this selection based on the material.
4. Select the manufacturing region for the part.
5. Select the use region of the part.
6. Select transportation modes.
7. Set the baseline. This baseline is the impact metrics values.
8. Iterate steps 2 to 6 with different materials and processes.
9. Select the best design based on the impact metrics values.

1.12 Case Studies

In this section, we use Dassualt Systems SolidWorks 2011 Sustainability Xpress to show how the impact metrics and the LCA concepts are used for sustainable design and analysis. We offer two cases: one for the design of a part (sprocket) and one for the design of an assembly (hex bolt-nut).

We have chosen the Sprocket part and the Bolt-nut assembly since they are widely used in building and joining several components to form an end product such as couplings, engines, or any machine design. Considering that these components are usually manufactured in bulk and account for significant use of natural resource, it is imperative to understand the resource use and emissions during development as well as their use of such components.

SolidWorks allows us to

1. Apply various materials and determine the material that causes the least impact to the environment.
2. Select from a wide range of manufacturing processes based on the material, manufacturing region, transportation and use.
3. Evaluate the design process throughout the life cycle of a product.

1.12.1 Case 1: Sprocket Sustainable Design

We evaluate an American National Standards Institute (ANSI) standard 14-tooth sprocket (shown in Figure 1.1 (a) and Figure 1.1 (b)), which is mainly used in chain-driven products such as automotive and pumps, to transmit rotary motion between shafts, where using gears are a misfit.

The sustainability analysis of the sprocket requires us to establish the environmental baseline first. The SolidWorks workflow to establish the baseline is

■ Evaluate (Ribbon Tab).
■ Select Sustainability or Tools (menu).
■ Sustainability Xpress.

The detailed steps are as follows:

Step 1: Select material, based on Class (Steel) and Name (Alloy Steel) criteria in the tool (Figure 1.2).
Step 2: Select the appropriate Manufacturing Process (CNC Milling) and Manufacturing Region (China, Asia) (Figure 1.3).
Step 3: To complete the baseline setting process, SolidWorks Sustainability tool requires the *transportation and use region* of the product to be specified. North America was selected as the "Use Region," considering it is one of the largest consumers of automobiles (Figure 1.4; as per CNBC's recent global survey for World's Largest Auto Markets - http://www.cnbc.com/id/44481705/World_s_10_Largest_Auto_Markets?slide=10).
Step 4: Baseline can be set using this icon, located at the bottom of the tool.

The Sustainability tool makes calculations and reports the results as per metrics. The values of the impact factors are based on every input considering all stages in the life cycle of the product being assessed.

The consolidated reports are generated in a pie chart format (Figure 1.5).

(a) Sprocket CAD model

14 Tooth Sprocket

(b) Sprocket engineering drawing

Figure 1.1   Sprocket CAD Model and Engineering Drawing.

The total environmental impact caused due to the ANSI standard 14-teeth sprocket is 0.51 kg CO_2, 5.72 MJ of energy, 2.92 × 10^{-3} kg SO_2 (equivalent) of sulfate, and 3.92 × 10^{-4} kg PO_4 (equivalent) of phosphate. SolidWorks also provides the flexibility to assess an impact metric at every stage of a life cycle, individually.

Figure 1.6(i) shows the carbon footprint. The emissions predominantly include carbon dioxide and greenhouse gases. Figure 1.6(ii) shows the energy consumed in procuring the iron (the material considered in this case is steel), manufacturing the sprocket, as well as disposing of it after its life. The energy consumed includes all nonrenewable energy resources and energy used in conversion such as hear, power, and so forth. Figure 1.6(iii) shows air acidification caused in the entire life cycle

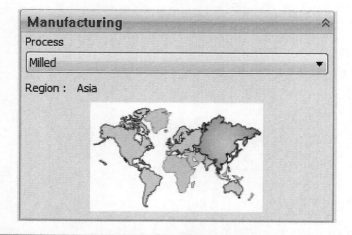

Figure 1.2 Step 1 of Establishing the Baseline.

Figure 1.3 Step 2 of Establishing the Baseline.

of a sprocket. Figure 1.6(iv) shows water eutrophication caused in the entire life cycle of a sprocket. The sum of an impact metric at every stage (as shown above) gives the total of each impact metric for the entire product life cycle.

A sustainability assessment and evaluation of the sprocket were made using the SolidWorks® sustainability tool by considering several design alternatives. Tables 1.2 through Table 1.4 show the result. In Table 1.2, we change the sprocket material from iron to aluminum while holding the other parameters constant. The impact metrics shown in the table indicate that using iron is a better sustainable design of the sprocket. Note that the water eutrophication for both materials is almost the same. Table 1.3 shows the effect of the manufacturing process. It shows that milling is much more environment-friendly than sand casting. Finally, Table 1.4 shows manufacturing in India is better for the environment than manufacturing in other

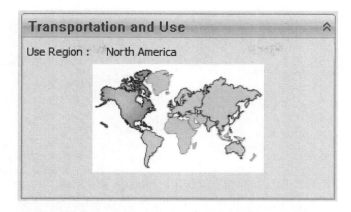

Figure 1.4   Step 3 of Establishing the Baseline.

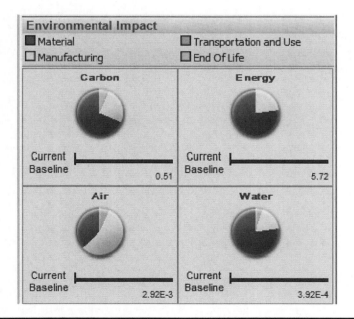

Figure 1.5   The Consolidated Report.

parts of Asia including China. This is due to the transportation distance. India is closer to the United States than is Asia.

The assessment of the six design alternatives is shown in graphical form in Figure 1.7, with designs (cases) shown along the X–axis and all four corresponding impact metrics shown along the Y–axis.

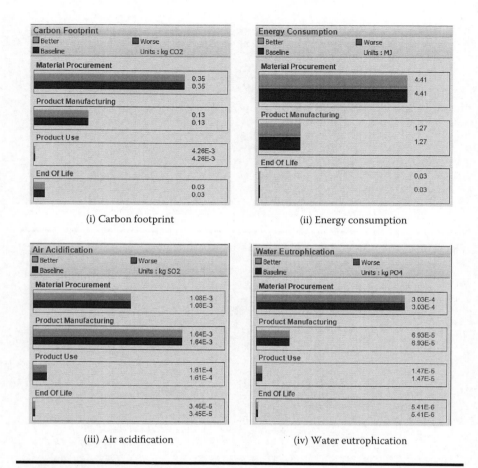

(i) Carbon footprint   (ii) Energy consumption

(iii) Air acidification   (iv) Water eutrophication

Figure 1.6   **Impact metrics of a sprocket.**

In Figure 1.7, A & B represent the effects due to change in material, C & D represent the effects due to change in manufacturing processes, and E & F represent the effects due to change in manufacturing region. Sustainability using the LCA approach helps in evaluating the environmental effects of a product holistically, that is, material through use of the product. Each case plots four impact metrics with the appropriate units along the Y–axis.

The results indicate assessment case D shows the highest levels of environmental impact, including all effecting factors such as carbon footprint, energy consumption, air acidification, and water eutrophication. Thus, the set of materials and processes followed throughout the life cycle in D should be avoided. Let us consider the comparison of D and C as they have the most life cycle stages in common. This comparison focuses on the manufacturing processes selected in each case. It indicates that the selection of manufacturing processes is a crucial parameter to preclude

Table 1.2 Material Effect

Case	Design Criteria	Carbon Footprint (kg CO_2)	Energy Consumption (MJ)	Air Acidification $\times 10^{-3}$ (kg SO_2)	Water Eutrophication $\times 10^{-4}$ (kg PO_4)
A	Material: Iron (Ductile Iron) Manufacturing Process: Sand Casting Manufacturing Region: Asia Use Region: North America	0.42	3.77	2.01	2.16
B	Material: Aluminum (1060 Aluminum Alloy) Manufacturing Process: Sand Casting Manufacturing Region: Asia Use Region: North America	0.80	9.71	5.33	1.99

Table 1.3 Manufacturing Process Effect

Case	Design Criteria	Carbon Footprint (kg CO_2)	Energy Consumption (MJ)	Air Acidification × 10^{-3} (kg SO_2)	Water Eutrophication × 10^{-4} (kg PO_4)
C	Material: Steel (Alloy Steel) Manufacturing Process: CNC Milling Manufacturing Region: Asia Use Region: North America	0.51	5.72	2.92	3.92
D	Material: Steel (Alloy Steel) Manufacturing Process: Machined Sand Casting Manufacturing Region: Asia Use Region: North America	1.12	12.21	9.41	7.20

Table 1.4 Manufacturing Region Effect

Case	Design Criteria	Carbon Footprint (kg CO_2)	Energy Consumption (MJ)	Air Acidification × 10^{-3} (kg SO_2)	Water Eutrophication × 10^{-4} (kg PO_4)
E	Material: Steel (ANSI 4130 Steel) Manufacturing Process: CNC Milling Manufacturing Region: Asia Use Region: North America	0.48	5.50	2.93	3.97
F	Material: Steel (ANSI 4130 Steel) Manufacturing Process: CNC Milling Manufacturing Region: India Use Region: North America	0.35	4.21	1.33	3.33

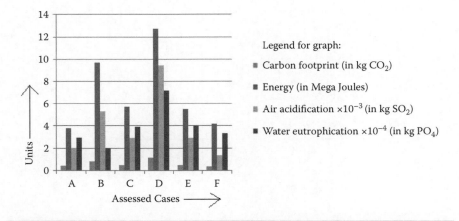

Figure 1.7   Impact metrics for the six sprocket designs.

such high levels of impacts to the environment. Case C results show a significantly low contribution to environmental impact. With this information, it can now be concluded that machined sand casting may be avoided for better environmental benefits for a particular PLC. The results also indicate that the assessment case A shows the best overall performance and the most optimized approach to develop a product like Sprockets. Case B follows the same set of processes and procedures except for a change in material of the product. This change has resulted in significant use of energy, and higher air acidification and carbon footprint. Cases E and F exhibit almost the same results, and the change in results is due to the change in manufacturing region.

1.12.2 Case 2

In Case 2, an assembly was evaluated, consisting of standard hex bolts and nuts with 1.992 inch length, 0.709 inch head diameter, 0.272 inch head thickness, and 0.21 inch pitch diameter (class 3A thread). Figure 1.8 shows the assembly CAD model.

This case study aims to analyze an assembly under two scenarios. In the first scenario, the bolts and nuts are manufactured at two different regions and directly transported to the use region as an assembly. In the second scenario, the hex bolts are transported to the region where the nuts are manufactured, and both are shipped to the use region as an assembly. Figure 1.9 shows the results.

Figure 1.9 shows the total environmental impact for the first design. This accounts for a total (for both the blot and the nut) of 1.68 kg CO_2 of carbon footprint, 19.67 MJ of energy, 10.34 × 10^{-3} kg SO_2 of air acidification, and 6.1 × 10^{-4} kg PO_4 of water eutrophication.

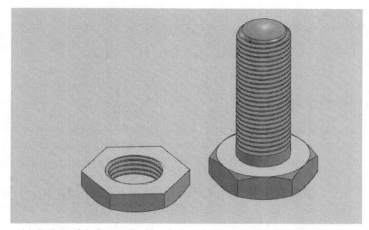

(a) CAD model of a standard hex bolt and nut, created in SolidWorks 2011

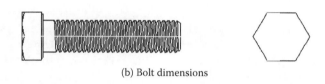

(b) Bolt dimensions

Figure 1.8 CAD model of a bolt-nut assembly.

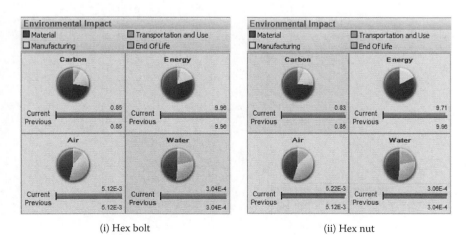

(i) Hex bolt   (ii) Hex nut

Figure 1.9 Impact metrics of a bolt-nut assembly.

In the second design, the environmental impact metrics are 1.242 kg of CO_2 carbon footprint, 14.61 MJ of energy, 7.69×10^{-3} kg SO_2 of air acidification, and 2.75×10^{-4} kg PO_4 of water eutrophication, indicating a better sustainable design.

1.13 Conclusion

This chapter covers the concepts and practice of the sustainable design philosophy. For the sustainable design notion to become mainstream, designers need easy-to-use design tools. The key for designers is to provide them with quantitative analysis to aid them in evaluating their design alternatives. As this chapter illustrates, designers must understand the basics of product life cycle and LCA concepts to be able to appreciate and relate to the results of sustainability tools such as SolidWorks Sustainability Xpress.

Authors

Pratheep Ayyamperumal holds a "Bachelor of Engineering" degree in Mechanical Engineering from College of Engineering Guindy, Anna University, Chennai, India. He has 6 years of Software Engineering experience in the development of CAD data interoperability and PLM tools with Triad Software Private Limited, Chennai, India. He also holds a "Master of Science" degree in Computer Systems Engineering from Northeastern University, Boston, MA, USA. His areas of interest include CAD/PLM and Software product development.

Ranjit Vinu is a graduate student at Northeastern University pursuing MS in Computer Systems Engineering. He has earned an undergraduate degree in Mechanical Engineering from Visvesvaraya Technological University. He has extensive experience in CAD/CAM tools, CAD Data Management and Product Lifecycle Management, having worked with PTC* for 3 years. His areas of interest include CAD Mathematics, Data Structures, and Parallel Programming.

Abe Zeid is a Professor with the Department of Mechanical and Industrial Engineering at Northeastern University. His research topics include the use of mobile agents to facilitate information access in manufacturing environments, developing XML-based algorithms for mass customization, and developing a Java-based and Web-based system for disassembly analysis. The system allows users to disassemble the components of a PC, and calculate the disassembly cost associated with each component. Dr. Zeid has written textbooks in the areas of CAD/CAM and the Internet/World Wide Web. He is an ASME Fellow.

Sagar Kamarthi received his PhD in Industrial Engineering from the Pennsylvania State University in 1994. His research interests are in prognostics and health management and mass customization. In the recent years he has been focusing his research on scalable nanomanufacturing and healthcare mass customization. He worked on several NSF-funded research projects and published his research contributions in reputed journals.

Tucker J. Marion is assistant professor in Northeastern University's College of Business, School of Technological Entrepreneurship. Dr. Marion's research is concentrated on product development, innovation, and entrepreneurship. He has held product development and manufacturing positions at large firms and has also started or co-founded several start-ups. He is a graduate of mechanical engineering from Bucknell University, holds a Master's from the University of Pennsylvania and Wharton School in technology management, and a PhD in industrial engineering from Penn State. His work has appeared in *Journal of Product Innovation Management, Design Studies, Research-Technology Management*, and *International Journal of Production Research*.

References

Cagno, E., Di Giulio, A., and P. Trucco. 2000. Planning the end-of-life management within the Product design process (pp. 299–308). *Proceedings of SPIE International Conference on Environmentally Conscious Manufacturing.* SPIE 4193: 299–308.

Campbell, N., D'Arcy, B., Frost, A., Novotny, V., and A. Sansom. 2006. *Diffuse Pollution: An Introduction to the Problems and Solutions.* London: IWA Publishing.

Field III, F.R., and J.R. Ehrenfeld. 1995. *Life Cycle Analysis: The Role of Evaluation and Strategy.* Washington: The National Academies Press.

Fu, Y., Diwekar, U., Young, D., and H. Cabezas. 2000. *Designing for Environment.* Process Design Tools for Environment. Boca Raton: Taylor & Francis.

Fussler, C., and P. James. 1996. *Driving Eco Innovation: A Breakthrough Discipline for Innovation and Sustainability.* London: Pitman.

Ginley, D.S., and D. Cahen. 2011. *Fundamentals of Materials for Energy and Environmental Sustainability.* Cambridge University Press.

Graedel, T.E., and B.R. Allenby. 1995. *Industrial Ecology.* New Jersey: Prentice Hall.

Kinross, J. 2002. Freshwater Acidification and Acid Rain. Napier University, UK.

Lambert, A.J.D., and S.M. Gupta. 2005. *Disassembly Modeling for Assembly, Maintenance, Reuse and Recycling.* Boca Raton: CRC Press.

Life Cycle Assessment Journals by SETAC North America (http://www.setac.org/node/32).

McLean, H., and L.B. Lave. 1998. A life cycle model of an automobile. *Environmental Science and Technology*, v32, n13, p. 322A–330A.

McDonough, W., and M. Braungart. 2002. *Cradle to Cradle: Remaking the Way We Make Things.* New York: North Point.

Manzini, E., and C. Vezzoli. 2002. *Product-Service Systems and Sustainability.* UNEP, Paris.

Manzini, E., and F. Jegou. 2003. *Sustainable Everyday. Scenarios of Urban Life.* Ed. Ambiente, Milan.

Marten, G.G. 2001. *Human Ecology: Basic Concepts for Sustainable Development.* London: Earthscan.

Murayama, T., Kagawa, K., and F. Oba. 1999. Computer-aided redesign for improving recyclability. *Proceedings of EcoDesign 1999. First IEEE International Symposium on Environmentally Conscious Design and Inverse Manufacturing*, p. 746–751.

Odum, H.T. 2007. *Environment, Power and Society*. West Sussex: Columbia University Press.

Pennington, D.W., Norris, G., Hoagland, T., and J.C. Bare. 2000. Environmental comparison metrics for life cycle impact assessment and process design. American Institute of Chemical Engineers.

Rees, W.E. 1992. Ecological footprints and appropriate carrying capacities: what urban economics leaves out. *Environ Urban* 4: 121–30.

Rosy, W., Chen, Navin-Chandra, D., and F.B. Prinz. 1993. Product design for recyclability: A cost benefit analysis model and its application. *Proceedings of the 1st IEEE International Symposium on Electronics and the Environment*: 178–83.

Schmidt-Bleek, F. 1995. The role of industrial products on the way towards sustainability. Wuppertal Institute.

Simon, M., Fogg, B., and F. Chambellant. 1992. Design for cost-effective disassembly, DDR/TR1. Manchester Metropolitan University, Manchester.

Simon, M., and T. Dowie. 1993. Quantitative assessment of design recyclability, DDR/TR8. Manchester Metropolitan University, Manchester.

Solow, R.M. 1992. An almost practical step towards sustainability, resources for the future. RFF.

United States EPA. 1995. Life-cycle impact assessment: A conceptual framework, key issues, and summary of existing methods. USEPA-452/R-95-002.

Vigon, B.W., Tolle, D.A., Cornaby, B.W., and H.C. Latham. 1992. Life-cycle assessment: Inventory Guidelines and Principles. U.S. Environmental Protection Agency.

Warhurst, A. 2002. Sustainability Indicators and Sustainability Performance Management. Mining, Minerals and Sustainable Development, World Business Council for Sustainable Development and International Institute for Environment and Development, Coventry, U.K.

Zussman, E., Kriwet, A., and G. Seliger. 1994. Disassembly oriented assessment methodology to support design for recycling. *CRIP* 43: 9–14.

Chapter 2

Cellular Manufacturing Systems

Yaowu Zhang

Research and Development Engineer, Powell Industries, Inc., Houston, Texas

Contents

2.1 Background of Cellular Manufacturing System

Even in today's modern world, manufacturing industries continue to play important roles in the prime wealth-producing activities, not only of industrialized nations but also of some developing countries. As worldwide manufacturing competition becomes more severe and intensifies, and affluent consumers' sense of value becomes more acute, the once popular and effective manufacturing systems of flow shop (mass) production and job shop production can no longer satisfy the customers' demand for multi-products and small-lot-size production. Moreover, individual manufacturing companies must prepare themselves well to be able to adopt multi-product, small-lot-size production in order to adapt themselves to rapid market movement, which is characterized by diversified, specialty-, and flexibility-oriented products, and a shorter product life (Ham 1985). Multiproduct, small-lot-size production is also called batch-type production or variety production. It is characterized by the following attributes: variety of production items, variety of production processes, complexity of productive capacity, uncertainty of outside conditions, difficulty of production planning and scheduling, dynamic situation of implementation, and control of production. Considering all of the previously stated difficulties, it is necessary to develop some useful theories for making practical, effective, and flexible multi-product, small-lot-size production systems. Consequently, a series of different manufacturing techniques and philosophies such as PBC (production batch control), JIT (just-in-time), MRP II (materials resource planning), and FMS (flexible manufacturing systems) have been developed and practiced in some industrialized countries. Among these techniques and philosophies, Group Technology (GT) has always been considered important in solving the problems facing current manufacturers today. GT should also be considered by manufacturers wanting to stay in a current productive market.

In its purest form, GT is defined as a manufacturing concept or philosophy that seeks to identify, exploit, and group similar parts and operation processes that take advantage of similarities in design, manufacturing, and implementation. GT aims to increase production efficiency through more effective design rationalization, data retrieval, manufacturing standardization, and manufacturing rationalization (Ham 1985). The basic concepts of GT have been practiced around the world for many years as part of good engineering practice and scientific management. Traditionally, in batch-type manufacturing for multi-products, small-lot-sized production, each part is treated as unique from the design through the manufacture (Ham 1985). However, with the aid of GT, by grouping similar parts into part families based on either their design attributes or their operation processes, several advantages are expected, which include, but are not limited to a mass production effect, the possibility of a flow-shop pattern on the process route, reduction of setup time and cost, simplification of material flow and handling, rationalization of design, and standardization of production processes (Ham 1985).

As one of the critical applications of GT in production systems, Cellular Manufacturing Systems (CMS) is a production process that involves processing a collection of part families in which all parts bear some similarities to the dedicated machine cells, each of which resides in a cluster located physically on the floor and which are functionally independent, dissimilar machines (Irani 1999). This definition indicates an important aspect of CMS: the parts that are grouped into part families, the machines that are formed into machine cells, and the part families that are assigned to the machine cells for processing. As a unique manufacturing concept, CMS differs from other manufacturing techniques and strategies such as flow shop and job shop. In a traditional flow shop manufacturing environment, high productivity and low flexibility are achieved. In the conventional job shop manufacturing environment, low productivity and high flexibility are obtained. As in the overlapping of flow shop and job shop, CMS seeks to deliver both high productivity and high flexibility.

2.2 Evolution of Manufacturing/Production Systems

Manufacturing/production systems are concerned with the conversion or transformation of "raw" materials into products or services, specifically, one-of-a-kind or multiple copies. The core point in manufacturing is to meet the challenges of producing quality products at low price and in a timely manner. Over the years, manufacturers and practitioners have constantly looked for ways to reach this point through technological innovation, machine tools replacement, computing technologies enhancement, and/or new management strategy. In most cases, researchers and practitioners focus on technology innovations.

Perhaps as early as the 6th century B.C., the Phoenicians were engaged in the production of bricks in large quantities (Wild 1972). Because the Phoenician system of manufacturing/production was achieved only through the massive use of labor, their production method differed completely from the ones that might be adopted today. In the modern sense of manufacturing/production, the Phoenicians did not practice one particular production principle.

Manufacturing/production could be traced back to the early decades of the 1900s when new approaches were being tried in manufacturing/production methodologies, as well as in scales of production (Wild 1972). The stimulus for this kind of manufacturing/production derived largely from new inventions and the increasing availability of mechanized or automated methodologies applied to manufacturing/production systems. The development of these methodologies depended on product design and the use of power sources. The use of water, and then steam power, accelerated the development of manufacturing/production mechanism. The developments of machine tools and other mechanical production equipment climaxed in the 18th and 19th centuries (Wild 1972). As a direct result of this enhanced manufacturing/production environment, flow shop (mass) production came into the world as an independent manufacturing/production methodology. The flow shop manufacturing system is characterized by large-scale and stable volume of production, which appeals to the manufacturer for its cost economics, production effectiveness, and throughput rate efficiency. The flow shop manufacturing system is also characterized by its big disadvantage of lacking flexibility, which requires a very high volume of products to justify the initial high investment and to overcome the difficulties in product or process design changes.

When the history of the flow shop manufacturing system is considered, the inventions of Kay, Arkwright, Hargreaves, and others in the period during and following the Industrial Revolution facilitated the emergence of the textile industry as a flow shop (mass) production industry (Wild 1972). Other inventions and developments of tools such as lathes, drilling machines, shapers, and drop forges facilitated the production of products in large quantities to higher standards of accuracy (Wild 1972). Notice, however, that these examples related to the mass production of single-piece items or components destined perhaps for use in the manufacture of more complex products (Wild 1972). The production of such items in large quantities is an important feature of mass production, since it involves the semicontinuous use of special-purpose equipment for the production of large quantities of items. In the interest of clarity and simplicity, this feature of mass production is referred to as quantity production (Wild 1972). A second aspect of mass production deals with the manufacture of more complex items. The most important feature of this type of production is production flow, which can be referred to as flow production. Generally, mass production products cannot be manufactured by one tool or a piece of equipment due to their complexity or composite features (Wild 1972). They normally require the cooperation of several facilities. Mass production of such items is a more recent development than quantity production. It depends on the use of the flow principles, that is, the continuous flow of the products through or past a series of production facilities (Wild 1972). Flow production is most easily achieved with products that flow naturally. In contrast, hard discrete items do not possess these attributes. As a result, considerable efforts have been given to the design of flow systems for product manufacturing. As indicated in the flow systems, flow is not the only difficulty to be solved prior to the mass production of discrete complex products. Interchangeability of parts is another critical factor to be considered in the flow systems (Wild 1972). There is no doubt that each product is produced based on a set of standards or stipulations. However, there probably will be different standards for different parts. Therefore, it is essential to ensure that each of the many different standards of accuracy be achieved for the parts interchangeability. Similarly, a high degree of accuracy for each complex item is highly desired in flow production systems, since that accuracy will be required for complementary parts to match and fit with each other at a subsequent production stage (Wild 1972). In summary, the general terms of mass production consist of two technologies, namely, the technology for variety production and the technology for flow production. In mass production, variety production precedes flow production.

However, the flow shop manufacturing system, once considered the virtue of a free economic society, has been changed (Ham 1985). Job shop, which is now considered another conventional manufacturing system, diverges sharply from flow shop. The major difference between the job shop and the flow shop lies in the high flexibility of the job shop. The differences between the job shop and the flow shop result in a wide range of processing requirements for general-purpose machines and special-purpose-specified machines that produce specific products in small lots. In a job shop manufacturing system, similar machines are grouped together to form a functional machine cell or process center. Any type of product may be routed through the system, visiting only the necessary machine cells or process centers on the basis of operational sequences. Because each product or batch of products has its own unique routing through its respective system, a job shop manufacturing system can produce different varieties of products and still maintain high machine utilization inside the machine cell. However, as is well known, the major disadvantage of this layout is its low productivity, high material handling costs, high setup costs, high work-in-process inventories, long delivery lead times, and difficulties in production control. This is partially due to the time the product spent traveling around inside the machine cell and waiting for loading on different machines. That is why, as the variety of parts increases and production volume (batch size) decreases, a job shop layout does not become the most suitable production environment.

To maximize the utilization of the flow shop and the job shop manufacturing systems advantages and to minimize their disadvantages, cellular manufacturing systems have been developed as the new manufacturing concepts of group technology applications to the manufacturing/production systems. As presented in the previous section, a cellular manufacturing system focuses on dividing a system into subsystems called manufacturing cells. In cellular manufacturing systems, families of similar parts (part families) are identified, and the associated machine cells are formed such that one or more part families can be processed within a single cell (Seifoddini 1988). The effective formation of manufacturing cells is a basic step in the development and implementation of cellular manufacturing.

Considering all the evidence presented thus far, it is apparent that flow shop, job shop, and cellular manufacturing are closely related to each other.

Many researchers have studied different manufacturing systems and their relationship with the volume and variety in various manufacturing situations. Ham et al. (1985) classified the different manufacturing systems by volume and variety.

Based upon an estimation by Groover (1980), batch production and job shop production constitutes an important portion of total manufacturing activity, since as much as 75% of all parts manufactured are in lot sizes of 50 pieces or less. However, the job shop is inefficient in terms of lead times and total production costs. The lead times and production costs are critical factors in the competitive market. Therefore, a different production type is needed in order to increase the efficiency of the manufacturing process. Cellular manufacturing is such a production type that is located between mass production and job shop production. In summary, cellular manufacturing attempts to provide the efficiency of flow shop production, while maintaining the flexibility of a job shop or a batch production (Vakharia 1986). Cellular manufacturing is but one critical application of group technology principles in the area of production systems. Before the cellular manufacturing system is discussed, group technology will be presented in brief.

2.3 Historical Development of Group Technology

Although some researchers who are familiar with group technology rightly acknowledge S. P. Mitrofanov of the USSR as the originator of this technique, those who have studied this subject in depth agree that Mitrofanov's creative developments were founded upon a premise postulated by Flanders and A. P. Sokolovski. The concept of Group Technology originated with Flanders (1925), who described using product-oriented departments at a machine company to manufacture standardized products with minimal transportation. Later, in 1937, A. P. Sololovski strongly recommended that parts bearing similar configurations and features—all other elements being the same—should be manufactured in the same way by a standardized technological process. In 1955, a hierarchical monocode was installed successfully at a manufacturing company. The so-called planned group technology installation was made the following year. It was decided to use the same classification in conjunction with group technology to resolve a serious batching problem in manufacturing. Drawing from 6,000 candidate parts in the same year, the first production unit (cell) of two lathes, one vertical drill, and one milling machine became operational, had excellent results. It was not until after 1966 that widespread attention was given to group technology as a recognized manufacturing technique. It was this year that S. P. Mitrofanov's book titled *Scientific Principles of Group Technology* was published. In his book, Mitrofanov found that considerable reductions in setting time, and therefore increases in capacity, could be achieved with lathes if similar parts were loaded on the machine one after the other. He also pointed out that major savings in tooling costs could be achieved by this process of loading parts.

Mitrofanov's major contributions are that he specified design simplification and process standardization as imperative prerequisites to an effective GT program. He specifically suggested a dual classification and coding system: one part for design simplification and the other for process standardization.

He continued to argue that the classification of technical operations based upon component shape, surfaces, and features afforded the best solution to this problem. He illustrated the role of classification as the basic problem or solution on which Group Technology rests.

It is important to notice that Mitrofanov, Zvonitsky, and others had different opinions on batching methods.

Mitrofanov referred to the unification of production process as consisting of two types, namely:

■ Standardization of the production process regardless of layout, process, and/or product
■ Group production method utilizing a committed cell of machines

Another way of saying this is that for least-cost batching, there are two additional methods as follows:

■ Scheduling by family for grouping for manufacture according to optimum standardized processes in process-oriented layouts
■ GT cells of optimum standardized production processes

As of today, group technology has become an effective and efficient manufacturing philosophy. The ultimate use of group technology could be classified into four major areas, namely, part design, process planning, production, and other areas. Group technology is one of the most important aspects in the design of cellular manufacturing.

2.4 Advantages of Cellular Manufacturing Systems

Because of worldwide use of cellular manufacturing systems, more and more advantages of its use have been identified and proved. These advantages can be classified into visible advantages and invisible advantages.

The visible advantages of manufacturing cells are due mainly to the proximity of all machines required to make a family of parts (Wemmerlov and Hyer 1989). This normally decreases the total traveling distance required by the batches of parts in the family. Correspondingly, this results in a decrease in the total cost to the manufacturer. Second, the lead time between successive operations is reduced such that the average work-in-process (WIP) levels of parts are also reduced. Third, setup time is reduced because parts in the part family that have identical or similar operation sequences can be sequenced consecutively throughout the manufacturing cell. Finally, operators at adjacent machines can provide feedback to each other about defective parts and thus make it easy to control the quality of the parts or assemblies. In summary, cellular manufacturing cells have the capacity to minimize the manufacturing throughput times of entire part families by reducing the material flow times, machine setup times, and lead times. Furthermore, the creation of manufacturing cells allows the decomposition of a large manufacturing system into a set of smaller and more manageable subsystems. The material flows of these subsystems can be significantly simplified.

Manufacturing cells also have several invisible advantages (Vakharia and Kaku 1993). First, the implementation of manufacturing cells can be regarded as a management philosophy that enforces the pursuit of successful teamwork, uses of state-of-the-art manufacturing technology, and promotes the competitiveness of the organization. Second, manufacturing cells can lead to a higher motivation and morale due to greater job satisfaction. Third, the conversion to a cellular system prepares the way of automation. With computer control for the entire CMS, the management and production activities can be accomplished in a more automatic and economical manner.

As observed, the foregoing visible and invisible advantages can be summarized as follows:

- Reduced throughput time
- Lower raw material inventories
- Lower WIP inventories
- Lower finished goods inventories
- Faster responses to market demand in terms of quantity and customization
- Lower costs of part traceability on the shop floor
- Less setup time
- More efficient material handling
- Improved output quality and less rework

- Higher productivity
- Improved job satisfaction for operators
- Increased job status visibility
- Simplified planning and control procedure
- Improved quality control
- Reduced lead time
- Improved operator efficiency and flexibility
- Reduced manufacturing overhead costs
- Improved product design
- On-time delivery
- Better equipment utilization
- Better use of skilled labor
- Job satisfaction
- Improved information flow
- Reduction in space needs
- Reduction in labor cost
- Increase in utilization of equipment now in the cells
- Reduction in quantity of equipment required to manufacture cell part
- Reduction response time to customer orders
- Reduction in move distances/move times
- Increase in manufacturing flexibility
- Reduction in unit cost
- Improvement in employee involvement

2.5 Disadvantages of Cellular Manufacturing Systems

Compared to the traditional functional and product layouts, the implementation of manufacturing cells could, however, have some disadvantages. These disadvantages could arise from the underlying design characteristics of cells and the limitations of the methodologies used to design and evaluate the cells. Those disadvantages are shown in part as high investment, less flexibility in handling changeovers, and imbalance of machine utilization.

First, cell implementation often leads to an increase in investment, as certain machines need to be duplicated to achieve independent cells in order to improve productivity. Second, it could be argued that a CMS is less flexible than a functional or process layout. A CMS may lack the ability for changeovers due to the nonavailability of equipment. Hence, it is generally true that a CMS fails to deal with long-term changes in product mix flexibility. Third, another potential disadvantage of manufacturing cells is the lower machine and labor utilization compared to a functional or process layout. To create completely independent cells, some machine types must be duplicated among several cells. It is not possible, however, to operate manufacturing systems where all machines and operators are equally utilized. Finally, when machine breakdown occurs, the production rate of a cell may be hindered because it lacks additional machines of similar function to replace the disabled machine.

The methodologies available for designing and evaluating manufacturing cells could also generate some disadvantages. First, a primary problem is the absence of a flexible yet simple and comprehensive cell formation method that can rapidly and cheaply identify manufacturing cells for a multiproduct facility. This is complicated by manufacturing environment changes. Next, collection and verification of data, selection of suitable data analysis methods, and interpreting or adapting these results can be time consuming. Finally, most cell formation methods assume that the population of parts and the manufacturing facilities used as input are stable over time. Unfortunately, several dynamic operational factors can influence the machine allocations, part family assignments, and layout for the cells (Wemmerlov and Hyer 1987). The cell and family compositions could change on a short- or long-term basis.

2.6 Assumptions in Cellular Manufacturing Systems

In cellular manufacturing systems, most of the assumptions are very typical and can be characterized as the following:

- The number of parts, machines, and operations on each part are collected and screened through the available information systems.
- There is no part setup time.
- The system considered is fully automatic.
- All machines are 100% reliable before they fail.
- One machine can only process one part at a time.
- Workload balancing in different manufacturing cells is not allowed.
- Any part should be grouped into a part family once the part goes into the production line.
- Part and machine grouping is based on their similarities with each other, respectively.
- If possible, machines are formed into machine cells such that all the operations of a part family are completed in a single cell. If this is not possible, then add additional machines to minimize part inter-cell traveling based on the given financial criteria.
- Operations necessary to process a part should not be split between cells or between machines in the same cell.
- For all the parts, there is at least one feasible cell in which all operations can be completed.
- Parts may have more than one feasible cell.

2.7 Major Topics in Cellular Manufacturing Systems

Although CMS has been studied and practiced widely by many researchers and practitioners, the issues, such as grouping parts into part families, forming machines into machine cells, and clustering part families into machine cells, are still under developing processes in both practical life and the theoretic world. This does not mean that nobody has ever yet created efficient mathematical models or heuristic algorithms for these issues. On the contrary, there exist some nice techniques that have the ability to improve productivity while maintaining production flexibilities. However, these techniques just address some perspectives of a partial production system. Consequently, the results yielded by the so-called effective techniques are incomplete or partial.

2.8 Issues in Cellular Manufacturing Systems

2.8.1 Varieties of Flexibilities

Flexibility is the ability a system relies on to adjust itself to internal and external changes. In cellular manufacturing (CM) systems, flexibility can be defined as the ability of the system to adjust its resources to any changes in relevant factors such as products, processes, loads, and machine failures. According to Choobineh (1988), a change is a disturbance of any magnitude in the condition under which the system was originally conceived and its parameters specified. Cell formation is a flexible, complex, multidimensional, and hard-to-capture problem, with broad implications and organizations, due to different issues related to both system structures and system operations. These issues may incorporate different elements and attributes of a production facility, such as machine, product, processing, operation, routing, failure, design, product-mix, and system. A variety of schemes for classifying these different flexibilities have been proposed in the literature. These schemes have classified flexibility on the basis of product or process components, system operations, aggregates (Sethi and Sethi 1990), and long- and short-term needs (Gupta and Goyal 1989). In surveys by Lim (1987), Slack (1987), and Abdel-Malek et al. (2000), the following types of flexibility seemed to be the most important.

2.8.1.1 Routing Flexibility

The routing flexibility of a manufacturing system is its ability to manufacture a product by alternative routes through the system. It is dependent on the external product attributes, planning and scheduling, and the internal assignment of the equipment. This kind of flexibility can usually be achieved by providing alternative machines: general-purpose machines and/or numerous material handling systems. Routing flexibility is very popular in manufacturing systems for reducing production time, as some of the machines have failed or there exists an exceptional part.

2.8.1.2 Machine Flexibility

This refers to the various types of operations that the machine can perform without requiring a prohibitive effort in switching from one operation to the other. Machine flexibility can be achieved in practice by providing automatic tool changes, using fixtures, and supplying sufficient part loading devices. Machine flexibility is a fundamental of a manufacturing system and a prerequisite for most other flexibilities.

2.8.1.3 Process Flexibility

The process flexibility of a manufacturing system refers to the set of product types that the system can produce without major setups.

2.8.1.4 Product Flexibility

The product flexibility is the ease with which new products can be added or substituted for existing products. A survey conducted by Abdel-Malek et al. (2000) shows that product flexibility plays an important role in a manufacturing system because it helps achieve product diversification and a family marketing strategy.

2.8.1.5 Volume Flexibility

The volume flexibility of a manufacturing system is its ability to be operated economically at different overall output levels. This arises when the customer demand is not certain due to a lack of precise forecasting or some other unpredictable circumstances.

2.8.1.6 Product-Part Mix Flexibility

Product-part mix flexibility is a valuable addition, since it is gaining popularity as companies expand their businesses to suit customers' needs for specialty-oriented products. Product-part mix flexibility can be defined as the ability of a manufacturing system to produce different combinations of products economically and effectively at a given capacity without incurring major setup costs or extended setup times. This flexibility varies from time to time, depending on production volume, scheduling, and customers' needs.

2.8.2 Varieties of Production Issues

Any manufacturing system involves many production issues from the order entry to the final product and shipping. As far as a cellular manufacturing system is concerned, production issues can be summarized as

- Part demand
- Machine capacity

- Cost of inter-cell and intra-cell movements
- Production cost
- Backordering cost
- Processing time
- Production volume rate
- Operation sequence
- Part/machine cost
- Inventory holding cost
- Design and manufacturing attributes
- Setup cost, inspection cost, and fixture and tooling costs
- Number of machines
- Work-in-process cost
- Number of parts
- Cell layout
- Machine layout within the cell
- Types of product, machine, part, and operation variety
- Machine breakdown rate

2.8.3 Varieties of Constraints

Methodologies, which address cellular manufacturing systems, have their own conditions to be true. Without those constraints, the methodologies might not work in the proper way. These constraints, which differ and stand out from those given in existing research papers, can be summarized as

- Number of machine cells in the system
- Machine cell size limitations
- Part family size limitations
- Number of part families
- Number of machines or machine types
- Number of parts or part types
- Number of inter-cell transportation
- Routing
- Layout
- Low profit level
- Low utilization level
- Machine capacity
- No negativity
- Production and inventory balance
- Production capacity
- Annual operating budget
- Tool or processing requirement of parts
- Machine load

- Buffer capacity
- Alternative processing time ratio

2.8.4 Varieties of Cell Formation Objectives

With almost no exceptions, businesses are trying to make profits to survive or to stay in the market. There are two ways to do so, namely, by improving production efficiency and lowering production costs. Here it is assumed that the firm has enough business. This is the main reason a company must prepare itself to produce multiple, diverse products in a short life cycle, and then execute efficient production changeovers (Abdel-Maleck et al. 2000). In order to achieve this goal, some companies apply a cellular manufacturing system to their production business. Their objectives are obvious and can be summarized as

- Minimization of inter-cell/intra-cell movements (MH)
- Minimization of cell workload variation
- Minimization of processing time/costs
- Minimization of production, inventory, back-order costs
- Maximization of the number of parts produced
- Maximization of machine utilization
- Minimization of setup costs and fixed machine costs
- Minimization of setup time or maximization of machine scheduling flexibility
- Maximization of cell utilization
- Minimization of duplicated machines
- Minimization of machine investment
- Minimization of lead time
- Maximization of the ratio of output to input (profit)
- Maximization of load balancing
- Minimization of exceptional elements
- Minimization of work-in-process (WIP)
- Minimization of bottleneck machines
- Maximization of the sum of similarities
- Minimization of cycle time
- Maximization of similarity or compatibility measures

2.8.5 Varieties of Layouts

As can be seen from the existing literature, there are three different layouts:

- Process layout for flow shops
- Functional layout for job shops
- Cellular layout for cellular manufacturing systems

2.8.6 *Varieties of Data Structures*

There are three different data structures from the existing research work. This can be summarized as

- Binary data structure
- Weighted data structure
- Fractional data structure

2.9 Cell Formation Techniques

Cell formation consists of identifying part families and machine cells. Different approaches have been developed to solve the cell formation problem. One involves classification and coding of parts based on their features, for example, diameter, shape, and material. Many classification and coding systems have been developed to form part families. The classification and coding systems help minimize the unnecessary variety of components by making the designer aware of similar parts. However, this approach is expensive and needs considerable effort to design and implement (Kusiak 1987). Because of these limitations, many techniques have been developed to solve the cell formation problem based on production flow analysis (PFA) introduced by Burbidge (1971). In PFA, machines required by parts are represented by a 0-1 machine-part matrix (part routing sequence). In this matrix, the element 0 means that part type "p" does not require machine type "m," while the element 1 means that part type "p" requires machine type "m." Since the cell formation problem is very important but NP-complete in nature (King and Nakornchai 1982), most of the procedures are heuristic in nature.

Several applications and survey papers on techniques for solving the CF problem have been published. These techniques include matrix arrangement, similarity coefficient, graph theory, mathematical programming, and heuristic methods. The optimal solution may be achieved in most of these methods; however, they have several drawbacks:

- A thorough review of literature shows that identifying part families and machine groups is usually achieved by optimizing an indirect index, for example, a similarity or dissimilarity index through the part machine incidence matrix without considering other production-related data.
- The majority of the models do not take into account the cost associated with inter-cell transportation and inter-cell movements.
- Many models neglect several important factors, e.g., part processing routings, part demand, part volume, part rejection rate, machine capacity, machine failure rate, etc.

- They require manual inspection or subjective decision to identify part families and machine groups.
- They require long computational times to solve large-scale problems.
- The number of manufacturing cells to be used is specified *a priori*. The upper limit on cell size is also imposed arbitrarily prior to clustering of part families and machine cells.

A number of procedures have been reported for partitioning the part machine matrix without considering alternative process plans and additional copies of the mechanism. Burbidge (1971) introduced production flow analysis for grouping parts and machines in the existing facility. McAuley (1972) suggested the similarity coefficient approach. Several researchers have used the idea of similarity in part and/or machine grouping, including the single linkage clustering method (McAuley 1972), average linkage clustering method (Mosier 1989), and MACE heuristic (Waghodekar and Sahu 1984). Matrix-based procedures use part machine matrix and identify clusters in the matrix by rearranging rows and columns. A few selected procedures include rank order clustering or ROC (King and Nakornchai 1982), bond energy algorithm or BEA (McCormick et al. 1972), MODROC (Chandrasekharan and Rajagopalan 1986), and ZODIAC (Chandrasekharan and Rajagopalan 1987). Miltenburg and Zhang (1991) provide a comparative study of nine well-known algorithms. In the above procedures, assigning additional copies of machines available is often treated subsequent to the cell formation (King and Nakornchai 1982). Also, the procedures developed to form cells are decoupled from the evaluation of the groupings obtained.

Kusiak (1987) proposed a p-median model for grouping parts and illustrated the importance of considering alternate process plans in improving the groupability. The objective of the model is to maximize the similarity between parts. The model requires the number of medians (p) or cells to be specified in advance. One has to experiment with different values of p before a desired grouping is discovered. Subsequently, Kusiak and Cho (1992) developed a branching algorithm. This algorithm requires representations of process plan similarities on a transition graph, then partitioning this graph to obtain part families. Viswanathan (1996) improved the original p-median model by proposing a new similarity coefficient formula that can take both positive and negative values and thus reveal the extent of similarity as well as dissimilarity between two parts. The above similarity-based methods identify only the part (process) families. Once part families are identified by the algorithm or model, the manufacturing cells are formed subsequently.

Kasilingam and Lashkari (1991) formulated a nonlinear 0-1 integer programming model for simultaneous grouping of parts and machines in the presence of alternate processing plans. The objective considered in the model is to maximize the compatibility indices between machines and parts, which are defined based on tooling requirements and processing times. However, this model assumes that the upper limits on the number of machines and parts in each cell are known.

In a flexible manufacturing system, consideration of tooling similarity is important, and this model is appropriate if a restriction on cell size can be imposed by the production constraints.

Three general methodologies realized in order to solve cell formation problems are listed in Table 2.1. Notice that all of these methods are time consuming and involve many data analyses by done properly trained professionals. Comprehensive reviews and classifications for cell formation methodologies are presented by Selim et al. (1998) and Offodile et al. (1994).

2.9.1 Visual-Inspection-Based Method

The visual-inspection-based method (also called the eyeballing method) is the simplest and least expensive method. It involves the classification of parts into part families by visualizing either the parts themselves or their drawings and arranging them into groups based on general criteria. This method is very limited in scope when dealing with a large number of parts. The accuracy of grouping obtained by this method is generally considered to be the least effective (Irani 1999).

2.9.2 Part Classification and Coding Method

Part classification and coding method is an essential tool for successful implementation of group technology in the cellular manufacturing system, yet this method is a highly time-consuming and complicated activity. Based on the similarity of part design or attribute, the method attempts to group parts into families. Design and/or attributes of a part are expressed in the format of numerical numbers or letters or a combination of numbers and letters. This code number dedicated to each part provides a compact and consistent description of the attributes of each part. Therefore, it serves as a basis for sorting and grouping the parts into families. The cell of each family is identified by matching the attributes of the parts in each family to machine capabilities and the available capacity on those machines. How to maintain and retrieve a balance between the information effectively becomes one of the most important factors in selecting a classification and coding system.

2.9.3 Product Flow Analysis

Production flow analysis is a method for forming part families and machine cells by analyzing the production process data listed in the operation sheets, that is, machine/workstations, operations, operation sequences, etc. This method does not require a classification and coding system, but it needs reliable, well-documented operation sheets. Therefore, a drawback of this method is that it assumes the accuracy of the existing operation sheet, with no consideration given to whether those process plans are up to date or optimal with regard to the existing mix of machines (Irani 1999).

Table 2.1   Cell Design/Formation Methodology

Visual Inspection-based Method		
Part Classification and Coding Method	Monocode (hierarchical)	
	Polycode (chain-type)	
	Hybrid (mixed)	
Product Flow Analysis	Mathematical Programming	P-median Formulation
		Linear Programming
		Nonlinear Programming
		Dynamic Programming
		Goal Programming
		Mixed Integer Mathematical Model
	Heuristic Technique	Heuristic Method
		Production Flow Analysis
		Simulating Annealing Algorithm
		Generic Algorithm
		Artificial Intelligence
		Similarity Coefficient Method
		Graph Theory
		Tabu Search
		Simulation
		Assignment Allocation Algorithm
		Neutral Network Method
		Sorting-Based Algorithm
		Nonhierarchical Heuristic
		Extended Cluster Identification
		Concurrent Formation
		Expert System

Table 2.1 (*Continued*)   Cell Design/Formation Methodology

Visual Inspection-Based Method		
		Multi-objective
		Matrix Formulation
		Branch and Fathoming Algorithm
		Non-agglomerative
		Agglomerative
		Axiomatic Approach
		p-Median Approach
		Quality Function Deployment
		Association Rule Induction
		Syntactic Pattern Recognition
		Sequential Modeling Approach
		Experimental Design

2.9.3.1 Mathematical Programming

As mathematical programming techniques show the functional relationship between variables through the use of mathematical symbols and operations, it is common to use this method in cell formation because it is capable of incorporating many variables or objectives in the design to achieve the optimal solution. Mathematical programming techniques can be classified into six categories: nonlinear programming techniques (NLP), linear programming techniques (LP), dynamic programming (DP), goal programming (GP), mixed integer mathematical model (MIMM), and assignment problem (AP).

2.9.3.2 p-Median Formulation

As an integer programming formulation of clustering problem, the p-median method was developed by Kusiak (1987) in order to obtain a diagonal or close-to-diagonal structure for the clustered matrix. The objective of p-median formation is to maximize the total sum of similarities between parts. Once the part families are formed, machines are allocated to the part families in order to form machine cells.

The main disadvantage of p-median formulation is that the number of medians (p value) or cells must be specified in advance. Therefore, it might be required to solve a given problem numerous times in order to find the optimal solution.

Another disadvantage of p-median formulation is that it is NP-hard. When the problem size gets bigger and more complex, the computational time will be much longer.

Viswanathan (1996) proposed a new measure of similarity coefficient in order to overcome the disadvantages of the original p-median method. The proposed measure can take both positive and negative values and therefore can identify the extent of similarity as well as dissimilarity between machines or parts. However, it cannot combine any production-related data into consideration except the part–machine incidence matrix.

2.9.3.3 Heuristic Techniques

When an optimal solution is hard to achieve or computational time is to be reduced significantly, heuristic techniques are employed instead of mathematical programming techniques. It can be defined as a decision procedure or rules that guide the search process in resolving a problem. It does not guarantee a global optimal solution. However, it ensures a local optimal solution in a reasonable computational time period.

2.9.3.4 Graph Partitioning Approaches

When considering the part–machine incidence matrix as a graph, graph partitioning approaches will be utilized. Nodes represent machines and parts, while arcs indicate the processing of the parts. One set of nodes represent parts, and the other set of nodes represent machines. The main goal of graph partitioning approaches is to obtain subgraphs from the machine–part graph to identify part families and machines cells.

2.9.3.5 Artificial Intelligence Approaches

Because of its efficiency in terms of computational time and the ability to capture design knowledge, AI techniques such as fuzzy logic, expert systems, pattern recognition, and neural networks have been used for cell formation.

2.9.3.6 Generic Algorithms

Generic algorithms are stochastic techniques based on the mechanism of natural genetics. Differing from other conventional techniques, they start with an initial set of random solution called population. Each individual in the population is called a chromosome, representing a solution to the problem at hand. The representation scheme of the chromosome determines how the problem is structured.

2.9.3.7 Simulated Annealing

Simulation annealing mimics the process of cooling a physical system slowly in order to reach a state of globally minimum potential energy. The stochastic nature of this algorithm allows it to escape a local minimum, explore the state space, and find an optional or near-optimal solution.

2.9.3.8 Simulation

Simulation studies of cellular manufacturing systems have appeared only in recent research. Simulation models provide an effective mechanism for simplifying the interpretation of a result from a well-planned experiment by explaining a selected system response as a function of relevant input factors.

2.9.3.9 Similarity Coefficient Method

The similarity coefficient method is the primary method in this study for identifying part families and machine cells.

Because similarity coefficients can incorporate production data other than just the binary part machine incidence matrix, a variety of similarity coefficients have been defined by many researchers. Usually, parts are grouped into part families, and machines are formed into machine cells subsequently based on the similarity coefficients between parts and between machines. Therefore, manufacturing cells will be formed by assigning part families into machine cells based on the objective function.

In the past, many researchers and practitioners have proposed various similarity coefficients between parts and between machines. Among the majority of those proposed similarity coefficients, it is clear that if the similarity coefficient is in a nominator-denominator format, the nominator is usually the common things. That is, both parts requiring the same machine in calculating the similarity coefficient between two parts or both machines can process the same part with the same routing in calculating the similarity coefficient between two machines. It is also clear that the denominator is the sum of both parts requiring processing on all machines minus both parts requiring the same machine in calculating the similarity coefficient between two parts. In the same way, the denominator will be the sum of both parts requiring processing on all machines minus both parts requiring the same machine at the same time when calculating the similarity coefficient between two machines.

The past research work also indicates that most of the similarity coefficients in manufacturing cell formation focus on a single production factor. Therefore, the majority of them are unable to incorporate various production data at the same time.

Most of the past similarity coefficients further indicate that they originated from the use of the binary part–machine incidence matrix. Thus, there are no comprehensive similarity coefficients between the parts and between the machines. As a result of this, it is essential to develop some flexible and effective similarity coefficients between the parts and between the machines with respect to not only the binary part–machine incidence matrix, but also the various production information, such as the part processing time, the part demand, the part volume, the number of operations of the part, the rejection rate of the part, the machine capacity, the machine capability, and the machine failure rate.

2.10 Manufacturing Cell Design/Formation Strategies

Even though there are many different cell design/formation methodologies, there are a fixed number of cell design/formation strategies. Through the research, it has been found that three different cell design/formation strategies are dominant in cellular manufacturing systems. The first one is the part family grouping solution strategy, in which the part families are formed first and the machines are then clustered into the machine cells based on the part families. The second one is the machine cell grouping solution strategy, in which the machine cells are formed first, and the parts are then clustered into the part families according to the machine cells. The last one is a simultaneous grouping of the part families and the machine cells solution strategy.

2.11 Performance Measurements

There is no doubt that the performance of cell formation methodologies will be measured to verify the success or failure of a cellular manufacturing system design with regard to the given criteria. The past research work shows that people focus on the measurements of

- Machine utilization
  - Machine utilization indicates the percentage of time that a machine is used in production.
- Cell utilization
  - Cell utilization can be defined as the sum of the machine utilization with respect to the number of machines in the cell.
- System utilization
  - It can be expressed as the maximum number of parts that can be processed by the specific CMS considered divided by the maximum number of parts in each period that can be processed by the total CMS configuration.
  - The system utilization can also be expressed as total sum of cell utilization divided by the number of cells.
- Grouping efficiency
  - Grouping efficiency is an aggregate measure, which takes both the number of exceptional elements and machine utilization into consideration.
- Weighted grouping efficiency

- Grouping efficacy
  - Grouping Efficacy (Γ) is defined by Kumar and Chandrasekharan (1990) as the number of operations to be the number of ones in the original part machine matrix.
- Weighted grouping efficacy
- Global efficiency

2.12 Summary

Severe global competition and customer awareness of various products are forcing manufacturers to seek more efficiency, more flexibilities, and more timely deliveries at a low cost if at all possible. As one of the most important applications of group technology in production systems, the cellular manufacturing system seeks to deliver high productivity and high flexibility in order to suit the diverse marketing needs. Cell formation of part families and machine cells is the critical element in designing an efficient cellular manufacturing system. It is a very complex problem and needs to be investigated and resolved through heuristic algorithms powered by programming.

Author

Yaowu Zhang holds a B.S. degree in Mechanical Engineering at Chongqing Jianzhu University in China in 1998. He was admitted to the Graduate School of Cullen College of Engineering at University of Houston as a Research Assistant in 2000. Two years later, he graduated with a M.S. degree in Industrial Engineering. While working as a full-time employee, he earned his Ph.D. degree in Industrial Engineering in 2010. His research areas include Costs Improvement, New Products Development, Cellular Manufacturing Systems, Group Technology, and Operations Research. Based on his outstanding achievements at Powell Industries, Inc., he was promoted several times to the current position of Corporate R&D Engineer. Prior to that, he held various positions as manufacturing engineer, cost specialist, programmer, special project manager, and production engineer. He is a longtime member of Houston Electrical League (HEL). He is also an active IEEE member.

References

Abdel-Malek, L., Das, S.K., and C. Wolf. 2000. Design and implementation of flexible manufacturing solutions in agile enterprises. *International Journal of Agile Management Systems* 2: 187–95.

Burbidge, J.L. 1971. *Production Planning,* London, Heinemann.

Chandrasekharan, M.P., and R. Rajagopalan. 1986. MODROC: An extension of rank order clustering for group technology. *International Journal of Production Research* 24: 1221–33.

Chandrasekharan, M.P., and R. Rajagopalan. 1987. ZODIAC—An algorithm for concurrent formation of part families and machine cells. *International Journal of Production Research* 25: 835–50.

Choobineh, F. 1988. A framework for the design of cellular manufacturing systems. *International Journal of Production Research* 26: 1161–72.

Flanders, R.E. 1925. Design, Manufacture, and Production Control of a Standard Machine, *Transaction of ASME* 46: 691–738.

Groover, M.P. 1980. *Automation, Production Systems, and Computer-Aided Manufacturing*, Englewood Cliffs, New Jersey, Prentice-Hall.

Gupta, Y.P. and S. Goyal. 1989. Flexibility of Manufacturing Systems: Concept and Measurement. *European Journal of Operation Research* 43: 119–135.

Ham, I. 1985. *Group Technology Applications to Production Management*, Boston-Dordrecht-Lancaster, Kluwer-Nijhoff Publishing.

Irani, S.A. 1999. *Handbook of Cellular Manufacturing Systems*. New York: John Wiley & Sons.

Kasilingam, R.G., and R.S. Lashkari, 1991. Cell formation in the presence of alternate process plans in flexible manufacturing systems. *Production Planning and Control* 2: 135–41.

King, J.R., and V. Nakornchai. 1982. Machine-component group formation in group technology: Review and extension. *International Journal of Production Research* 20: 117–33.

Kumar, C.S., and M.P. Chandrasekharan. 1990. Grouping efficacy: A quantitative criterion for goodness of block diagonal forms of binary matrices in group technology. *International Journal of Production Research* 28: 233–43.

Kusiak, A. 1987. The generalized group technology concept. *International Journal of Production Research* 25: 561–69.

Kusiak, A., and M. Cho. 1992. Similarity coefficient algorithms for solving the group technology problem. *International Journal of Production Research* 30: 2633–46.

Lim, S.H. 1987. Flexible Manufacturing Systems and Manufacturing Flexibility in the United Kingdom. *International Journal of Operations and Production Management* 7: 44–54.

McAuley, J. 1972. Machine grouping for efficient production. *The Production Engineer* 52: 53–57.

McCormick, W.T., Schweitzer, P.J., and T.W. White. 1972. Problem Decomposition and Data Reorganization by a Clustering Technique. *Operations Research* 20: 993–1009.

Miltenburg, M., and W. Zhang. 1991. A comparative evaluation of nine well known algorithms for solving the cell formation problem in group technology. *Journal of Operations Management* 10: 44–72.

Mosier, C.T. 1989. An experiment investigating the application of clustering procedures and similarity coefficients to the GT machine cell formation problem. *International Journal of Production Research* 27: 1811–35.

Offodile, O.F., Mehrez, A., and J. Grznar. 1994. Cellular manufacturing: A taxonomic review framework. *Journal of Manufacturing Systems* 13: 196–220.

Seifoddini, H. 1988. Comparison between single linkage and average linkage clustering techniques in forming machine cells. *Computers and Industrial Engineering* 15: 210–16.

Selim, H.M., Askin, R.G., and A.J. Vakharia. 1998. Cell formation in group technology: Review, evaluation, and directions for future research. *Computers and Industrial Engineering* 34: 3–20.

Sethi, A.K., and S.P. Sethi. 1990. Flexibility in manufacturing: A survey. *The International Journal of Flexible Manufacturing Systems* 2: 289–328.

Slack, N. 1987. The Flexibility of Manufacturing Systems. *International Journal of Operations and Production Management* 7: 35–45.

Vakharia, A.J. 1986. Methods of cells formation in group technology: A framework for evaluation. *Journal of Operations Management* 6: 257–71.

Vakharia, A.J., and B.K. Kaku. 1993. Redesigning a cellular manufacturing system to handle long-term demand changes: A methodology and investigation. *Decision Sciences* 24: 909–30.

Viswanathan, S. 1996. A new approach for solving the p-median problem in group technology. *International Journal of Production Research* 34: 2691–700.

Waghodekar, P.H., and S. Sahu. 1984. Machine-component cell formation in group technology: MACE. *International Journal of Production Research* 22: 937–48.

Wemmerlov, U., and N.L. Hyer. 1989. Cellular manufacturing in the US industry: A survey of users. *International Journal of Production Research* 27: 1511–30.

Wemmerlov, U., and N.L. Hyer. 1987. Research issues in cellular manufacturing. *International Journal of Production Research* 25: 413–31.

Wild, R. 1972. Mass-Production Management. *The Design and Operation of Production Flow-line Systems*, New York, John Wiley & Sons.

Chapter 3

An Overview of Computer-Aided Design

Ali K. Kamrani, PhD, PE
Industrial Engineering Department, University of Houston, Houston, Texas

Contents

Abstract

This chapter will present an overview of computer-aided design (CAD) and feature representation methodologies. It includes *feature's definition, Wireframe Modeling, Surface Modeling, Boundary Representation (B-rep), Constructive Solid geometry (CGS), and definition of interacting features*. CAD and the supporting methods are used to facilitate the engineering design process.

3.1 Introduction

The nonintegrated approach to design and manufacturing is widely recognized as a major contributor to increased product development costs. Parts are typically represented using *wireframe, boundary representation (B-rep)*, and *constructive solid geometry (CSG)*. Manufacturing information needed for higher-level applications, such as process planning, could not be extracted directly from these models, and therefore feature-based modeling has proved to be an effective and time-saving approach for product design. Features provide a better way to manage the design at higher level of description, although the traditional CAD descriptions that characterize an object in terms of mathematical surfaces or volumes are still a major part of this process.

The term *feature* can have different meanings depending on the specific domain. For example, in design it may refer to a notch section, while in manufacturing it refers to slots, holes, and pockets. In the inspection terminology, it is used as a datum or reference on a designed object (Chandra and Ghosh 1999). In general, classification of features is totally dependent on the application. It is very difficult to make an application-independent classification of features. There are many definitions in the literature for the term *feature* (Sheu and Lin 1993; Somashekar and Michael 1995; Case and Harun 2000; Ciurana and Romeu 2003; Santosh et al. 2003), some of which are as follows:

- "A feature is any entity used in reasoning about the design, engineering, or manufacturing of a product" (Sreevalsan and Shah 1992).
- "A geometric form or entity whose presence or dimensions are required to perform at least one CIM function and whose availability as a primitive permits the design process to occur" (Luby et al. 1986; Devireddy and Ghosh 1999).
- "A region of interest on the surface of a part" (Pratt and Wilson 1985).
- "A geometric form or entity that is used in reasoning in one or more designs or manufacturing activities" (Dixon and Cunningham 1988).

■ "A parametric shape associated with such attributes as its intrinsic geometric parameters-length, width, and depth as well as position, orientation, geometric tolerances, material properties, references to other features" (Mantyla et al. 1996).

Although many different definitions have been given, the basic idea underlying these definitions is that features represent the engineering meaning of the geometry of a part, assembly, or other manufacturing activity.

3.2 Geometric Data Format

Geometric modeling can be defined as a group of methods which are used to define the shape and other geometric characteristics of an object. Basically, the methods of geometric modeling are a synthesis of techniques from many fields such as analytic and descriptive geometry, topology, set theory, numerical analysis, vector calculus, and matrix methods.

Geometric modeling is a necessary element of computer-aided design (CAD) and has rapidly evolved due to increased computing power technology. Geometric modeling systems represent and manipulate three-dimensional objects in an unambiguous way. The main functions of the geometric modeling system are

■ Interactive design of a new solid object.
■ Modification and manipulation of the solid object using geometric operators such as rotation, translation, and Boolean operations.
■ Analysis of the properties of solid objects.
■ Effective and easy way of displaying the objects.

Currently, geometric modeling is required in both the fields of computer-aided design (CAD) and computer-aided manufacturing (CAM). Using CAD and geometric modeling tools, a designer can provide a complete and unambiguous model so that the manufacturing engineers can develop an accurate process plan and tool path of the numerical control machine, and perform an inspection of the final product. Therefore, for manufacturing applications, geometric modeling should be able to represent and manipulate the three-dimensional solid objects in a simple and non-ambiguous way (Yong and Tang 2000). Three different types of geometric modeling techniques are currently available. These are: *wireframe*, *surface*, and *solid modeling*.

3.3 Wireframe Modeling

The result of the reverse engineering process is typically a wireframe model that is ambiguous and lacks solidity. The generated wireframe graphic does not clearly define the surface information nor does it give the volume of the various regions

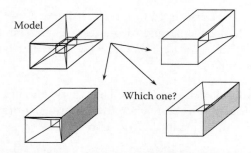

Figure 3.1   Wireframe representation model. (From Chang et al. 1990.)

of the part (Figure 3.1). Only points, lines, and curves are illustrated using wireframe modeling approach, and the model cannot determine the interior or exterior shapes of any surface within the object. Although the wireframe representation is not considered to be the major format for feature representation, it is still popular in design models for three main reasons: (1) It requires less expensive computing devices, (2) this method of modeling needs less computational effort, and it is extensively used in computer graphics such as displays, and (3) some models require interpretations from a paper drawing or physical part by optical equipment (camera). The information received is only in the form of wireframe data. This procedure is one of the important steps in the reengineering process.

3.4 Surface Modeling

Surface modeling represents a 3D object by describing part surfaces, but the interior properties are not defined. This approach can generate a picture equivalent to those of solid modeling by facilitating hidden line removal and shading effects (Kobbelt et al. 2001). Although this approach can automatically estimate the volume property from the surface definitions, the mass property and other important attributes for manufacturing cannot be extracted from this model because of undefined interior attributes. The section cut of an object using this approach can only reveal the edge definitions of the cut surface. The surface modeling approach can represent a more accurate surface definition than an equivalent solid model. Hence, this model is useful for the detection of the NC tool path. Three types of surfaces can be generated using this design approach (Chasen 1986):

■ Ruled or extruded surfaces
■ Surfaces of revolution
■ Sculptured surfaces

The extruded surfaces are typically formed by extruding one 2D closed curve. This technique is useful for the design of airplane wings. The surface of evolution can be created by rotating the 2D curve about an axis. For example, this method is useful in creating a symmetrical object such as a cylindrical part. The sculptured surface is used to create more complex surfaces such as a ship's hull or an automobile's fender. There are several techniques for creating 3D surfaces such as polygon meshes and parametric cubic patches. Even though the algorithm for polygon meshes is simple, the major disadvantage is that it represents an approximate curve using polygon meshes. As a result, the curve is not accurate, and an exact definition is not possible. Parametric cubic patches identify the coordinates of points on a curved surface using three parametric equations. Each bi-cubic patch is constructed from the corresponding curve. There are three typical types: *Ferguson's, Bezier,* and *the B-spline methods* (Wang and Wysk 1998).

3.5 Ferguson's Curve

Ferguson's curve is defined by two end points and two tangents at the end points as shown in Figure 3.2.

Ferguson's curve is defined by the following equation (Zeid 2007):

$$r(t) = \begin{bmatrix} P_0 & P_1 & T_0 & T_1 \end{bmatrix} \begin{bmatrix} 2 & -3 & 0 & 1 \\ -2 & 3 & 0 & 0 \\ 1 & -2 & 1 & 0 \\ 1 & -1 & 0 & 0 \end{bmatrix} \begin{bmatrix} t^3 \\ t^2 \\ t \\ 1 \end{bmatrix} \quad 0 \leq t \leq 1 \qquad (3.1)$$

The end points and the tangents at the end points are explicitly defined. Consequently, it can easily design a continuous and smooth curve segments. However, it is difficult to relate the tangent vectors to the overall shape of the curves.

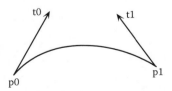

Figure 3.2   A Ferguson's curve segment.

3.6 Bezier's Curve

One solution to the problem of generating a smooth curve was developed in the late 1960s by two French automobile engineers, Pierre Bezier, who worked for Renault automobile company, and P. de Casteljau, who worked for Citroen. Originally, the solutions were considered industrial secrets, but Bezier's work was eventually published. The curves that result using Bezier's method are called Bezier curves. This method allows the designer to efficiently record information about shapes and quickly stretch, rotate, and distort these shapes. Bezier curves are commonly used in computer-aided design work and in most computer drawing software. This method is used to specify the shapes of letters of the alphabet in different fonts. By using this method, a computer and a laser printer can have many different fonts in many different sizes without using a large amount of memory. Bezier's curve $B(t)$ is defined for the four points P_0, P_1, P_2, and P_3 as shown in Figure 3.3 (Chang and Wysk 1984).

$$B(t) = \begin{bmatrix} P_3 & P_2 & P_1 & P_0 \end{bmatrix} \begin{bmatrix} 1 & 0 & 0 & 0 \\ -3 & 3 & 0 & 0 \\ 3 & -6 & 3 & 0 \\ -1 & 3 & -3 & 1 \end{bmatrix} \begin{bmatrix} t^3 \\ t^2 \\ t \\ 1 \end{bmatrix} \quad 0 \le t \le 1 \quad (3.2)$$

P_0, P_1, P_2, and P_3 are called *control points* for Bezier's curve. Bezier curves have a number of properties that make them particularly useful for design:

■ $B(0) = P_0$ and $B(l) = P_3$. In this scenario, the Bezier curve passes through points P_0 and P_3. This property guarantees that $B(t)$ goes through the specified points. This property guarantees that the control values of the Bezier curves at their end points can be controlled by choosing the appropriate values for P_0 and P_3.

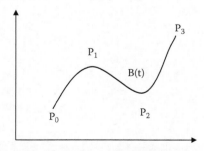

Figure 3.3   Bezier's curve $B(t)$.

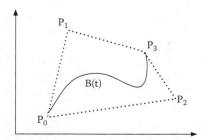

Figure 3.4   Property 4 of Bezier's curve.

- $B(t)$ is a cubic polynomial. This property guarantees that $B(t)$ is continuous and differentiable at each point. It also guarantees that the graph of $B(t)$ does not distort between control points.
- $B'(0)$ = slope of the line segment from P_0 to P_1: $B'(1)$ = slope of the line segment from P_2 to P_3. This property means that it is impossible to match the ending slope of one curve with the starting slope of the next curve for a smooth connection.
- For $0 \leq t \leq 1$, the graph of $B(t)$ is in the region that the corner points are the control points. Visually, this property means that if a rubber band is placed around the four control points of P_0, P_1, P_2, and P_3 as shown in Figure 3.4, then the graph of $B(t)$ will be inside the rubber-banded region. This is an important property of Bezier curves since it allows the graph of $B(t)$ to remain within the domain of the four control points.

3.7 B-spline Curve

The B-spline curve (nonuniform) is a general case of Bezier's curve. It is usually defined in a Cox–deBoor recursive function, as shown below (Chang and Wysk 1984):

$$r(t) = \sum_{i=0}^{L} N_i^n(t) p_i \quad t_i \leq t \leq t_{i+1}, \tag{3.3}$$

where

$$N_i^n(t) = \frac{(t - t_i)}{(t_{i+n-1} - t_i)} N_i^{n-1}(t) + \frac{(t_{i+n} - t)}{(t_{i+n} - t_{i+1})} N_{i+1}^{n-1}(t), \tag{3.4}$$

$$N_i^1(t) = \begin{cases} 1, & t \in [t_i, t_{i+1}] \text{ and } t_i < t_{i+1} \\ 0, & \text{otherwise} \end{cases},$$

L = number of control points and n = degree of the curve.

The Bezier and B-spline methods have many common advantages.

- Control points can be adjusted in a predictable way, making them ideal for use in an interactive CAD environment.
- Local control of the curve shape is possible. In this case, the Bezier method has a disadvantage because if the order of the polynomial is increased by adding more control points when more control of shape is needed, the order of the polynomial does not change in the B-spline method.

3.8 Solid Modeling

Solid modeling is one of the most comprehensive methods for design. This method overcomes the drawbacks of both wireframe and surface models by providing a comprehensive and complete definition of the three-dimensional object. This method recognizes that the solid object as a volumetric description, and it can also operate on the surface and edge definitions of an object. The mass, volume, surface, and other important engineering properties can be derived using this method.

The origins of solid modeling are traced back to the key technological inventions of the 1950s, which was computer graphics. This spawned the developments of computer-based geometric systems to aid in the description of an object's geometry, which is the main activity in the design and manufacture of mechanical parts. This led to research into the development of Computer-Aided Design and Computer-Aided Manufacturing (CAD/CAM). Preliminary systems used electronic drafting and wireframe models to represent the shape of three-dimensional objects. Subsequent systems developed in the 1960s used polygonal and surface-based models, which were utilized for a variety of applications in aerospace, marine, and automotive industries.

Until the 1970s, these models were used in a broad manner. They were merely a collection of lower-dimensional entities (polygons, surfaces, lines, curves, and points) put together in an unstructured manner to represent a real object. The developments in CAD/CAM led to the crucial questions about the uniqueness and the validity of these models, issues that until then were unimportant from the point of view of computer graphics and its applications. In the late 1970s, these issues were resolved by the Production Automation Project at the University of Rochester, where the term *solid modeling* was coined (Reqicha and Voelcker 1982). This group developed new mathematical models for representing solids. They also identified the mathematical operations that could be used to manipulate these models.

In the 1980s, several solid modeling systems were developed and used in the commercial CAD/CAM world, including the automobile, aerospace, and manufacturing industries. Moreover, many advanced CAD/CAM applications of solid modeling have emerged, such as feature- and constraint-based modeling, automatic mesh generation for finite element analysis, assembly planning, including interference checking, and higher-dimensional modeling for robotics and collision avoidance, tolerance modeling, and automation of process planning tasks. Currently, solid modeling techniques have gained importance in the industry and are also actively pursued as a research field in academic institutions. Figure 3.5 lists the evolution cycle of this critical technology.

Solid modeling provides a framework to model and represents an object's shape in the computer, and to perform operations. In addition, a group of application-independent geometric tools and algorithms is provided that can be used to query/analyze the model to obtain unambiguous results. These tools can be used or combined with other application-specific tools to perform the required task. The issues related to data structures and geometric algorithms, their efficiency, reliability, and robustness also form an important aspect of solid modeling.

The academic effort in solid modeling utilizes several disciplines in many applications, including algebraic geometry and topology, differential geometry and topology, combinatorial topology, computer science, and numerical analysis. Another well-established and closely related field is *Computer-Aided Geometric Design* (CAGD), which concentrates on developing techniques for free-form surface design used to model curves and surfaces (Natekar et al. 2004). Six basic types of solid modeling representations (Mantyla 1988; Wang and Wysk 1998) exist.

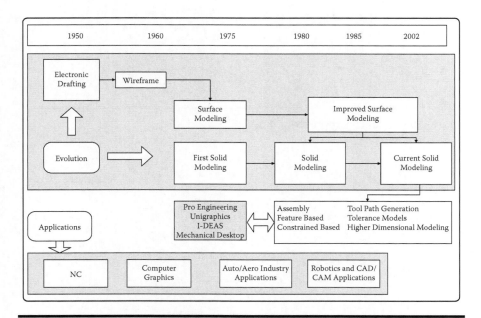

Figure 3.5   Solid modeling technology evolutions.

3.8.1 Boundary Representation (B-rep)

The object is represented by the means of the bounded faces that enclose it. Each face is represented by its bounding edges and vertices (Figure 3.6).

3.8.2 Constructive Solid Geometry (CSG)

The object is represented by an ordered tree of the Boolean operations. Nonterminal nodes represent Boolean operators such as union, intersection, or difference. Terminal nodes represent a primitive solid and transformation data (Figure 3.7).

3.8.3 Cell Decomposition

The object is represented as the sum or union of a set of cells into which it is divided. The disjoint cells can be of any shape and size. This representation technique is the base of finite element modeling (Figure 3.8).

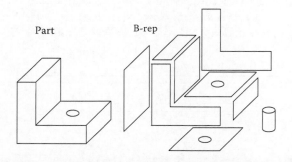

Figure 3.6   Boundary representation.

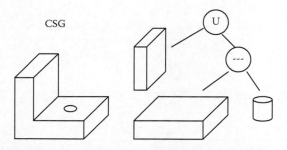

Figure 3.7   Solid constructive geometry.

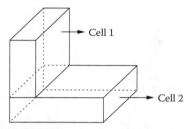

Figure 3.8   Cell decomposition.

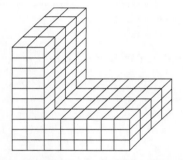

Figure 3.9   Spatial occupancy enumeration.

3.8.4 Spatial Occupancy Enumeration

The object is represented by a list of the cubical disjoint spatial cells that it occupies. This is the special case of the cell decomposition where the shape of the cells is cubical (Figure 3.9).

3.8.5 Primitive Instancing

The object is represented by a set of solid primitives such as cube, cylinder, cone, and so forth. Each primitive is usually defined parametrically and located in space (Requicha 1980; Requicha and Chan 1986).

3.8.6 Sweeping

The object is represented by moving a curve or a surface along some paths (Figure 3.10). This method is useful to model constant cross-sectional parts and symmetrical parts (Taylor 1992).

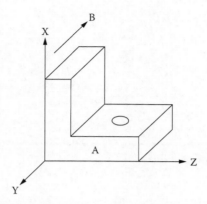

Figure 3.10   Translational sweeping.

3.9 Boundary Representation (B-rep)

Boundary representation (B-rep) is one of the methods that are extensively used to create a solid model of a physical object and associated geometric data model (Sheu and Lin 1993). Boundary representation describes the geometry of an object in terms of its boundaries, namely, the vertices, edges, and surfaces that represent entities of two dimensions, one dimension, and zero dimension, respectively (Chang 1990). In order to represent a solid object by its surfaces, the orientation of each surface must be defined as the interior or exterior of the object. Typically, the inside is the material part, and the outside is the void space. In a boundary representation, a face must satisfy the following conditions:

- A finite number of faces define the boundary of the solid.
- The face of a solid is a subset of the solid's boundary.
- The union of all the faces defines the boundary.
- The face itself is a limited region or subset of the more extensive surface.
- A face must have a finite area and is dimensionally homogeneous.

A solid is bounded by surfaces, and one can then define a solid by a set of faces. Consequently, the enclosed volume of the object can be completely defined. Figure 3.11 shows the topology of the solid model that presents the object as a set of faces. Each face is bounded by edges, and each edge is bounded by vertices. To separate the points that are inside or outside of the object, the direction of the surface normal is used to attach the face. This information is typically encoded by numbering the edges in a sequence such that the right-hand rule defines the vector that points outward from the object.

Geometrical data includes the coordinates of the vertices and transformation (translation and rotation), metric information, such as distances, angles, area,

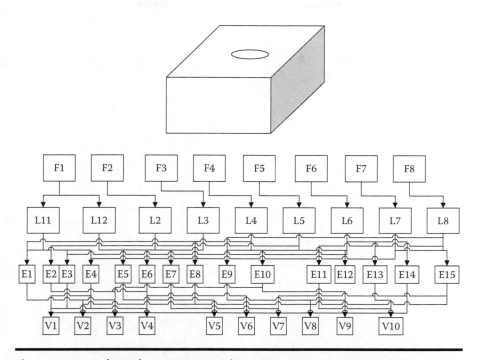

Figure 3.11   A boundary representation. (From Wang, H.P., and R.A. Wysk. 1998. *Computer Aided Manufacturing*. New Jersey: Prentice Hall.)

and volumes. In order to use boundary representation, the topology relations should be stated between each set of surfaces in order to identify the surfaces in order to identify the shape and volume of the object.

Parts can be classified as either *polyhedral* or *curved* objects. A polyhedral object is presented by planer faces connected with straight edges that, in turn, are connected at vertices. A boundary representation model is not limited to a planer surface. Different types of surfaces geometries can be described by different B-rep models that approximate curved surfaces as a combination of planer surfaces.

The validity of the B-rep model can be validated using Euler's formula (Kang and Nnaji 1993). Each edge in a topological valid B-rep is always adjacent to exactly two faces and is terminated by two vertices. Euler's formula states that for any topological valid solid, the number of vertices in addition to the number of faces is equal to the number of edges plus two

$$V + F = E + 2 \qquad (3.5)$$

where
   V = number of vertices,
   F = number of faces,
   E = number of edges.

Topological correctness (all necessary geometric elements are presented and connected properly) does not guarantee a valid solid. If the geometry is altered, the topological model no longer represents the actual shape of the designed part. The geometric and topological definitions of the model may become contradictory, which result in an invalid object. Therefore, Euler's formula does not always give a valid object. For example, Figure 3.12 shows geometrical and topological definitions that are conflicting and therefore result in an invalid object. Figures 3.12(a) and 3.12(b) show the surface modification that results in a valid object. On the other hand, Figure 3.12(c) shows a surface modification that results in an invalid object.

The result when two B-rep blocks are combined together is shown in Figure 3.13. Originally, there are 8 vertices on each block, and two are inside the combined object. Six new vertices are created from the combined object. The total number of vertices (V) is now 20. The basic two blocks have 24 edges, and all exist with the object. Six more edges are created by the union operation. The total number

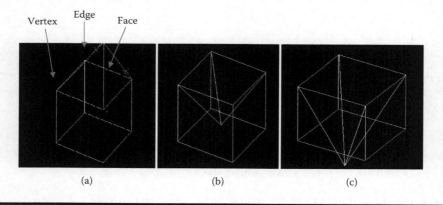

Figure 3.12   Valid and invalid objects.

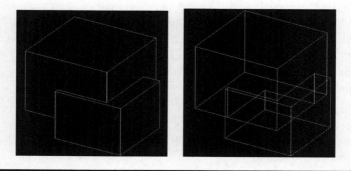

Figure 3.13   Two cubes combined together.

of edges (E) is 30 (24 + 6), and the total number of faces is 12. Therefore, after applying Euler's formula (20 + 12 = 30 + 2), it can be concluded that the resulting object has the right number of topological elements.

3.10 Constructive Solid Geometry (CSG)

The CSG is a modeling method that defines complex solids as compositions of simple solid primitives (Mortensen 1985). In CAD systems, primitives can be either from a fixed set of predefined primitives built in the CAD system or defined primitives created specially by users (Chang 1990). For example, the standard solid primitives shown in Figure 3.14 are block, cylinder, sphere, torus, wedge, and cone. In the CSG model representation, the part design is represented by an ordered binary tree. The ordered binary tree consists of nodes, where these nodes are either terminal or nonterminal (Waco and Kim 1994) as shown in Figure 3.15. Terminal

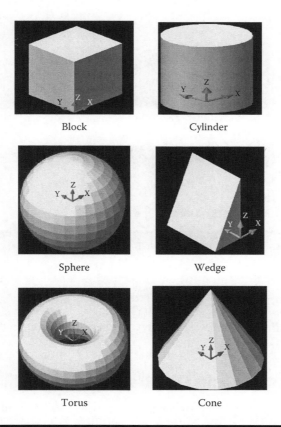

Block   Cylinder

Sphere   Wedge

Torus   Cone

Figure 3.14   Solid primitives.

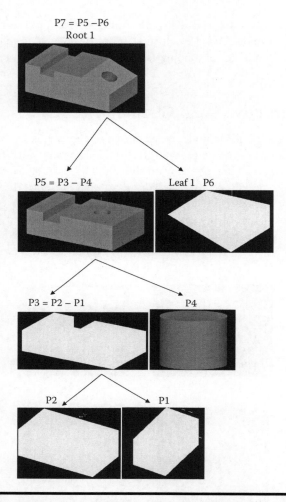

Figure 3.15 Constructive solid geometry (CSG) tree for a sample part.

nodes (leaf) can be either primitive or primitive with transformations. Nonterminal nodes (roots) are either Boolean operators (union U, difference -, and intersection ∩) or solid object motions (translation and/or rotation). Each nonterminal node represents a combination of two solid primitives in the tree (Mortensen 1985).

3.11 Advantages and Disadvantages of (CSG) and B-rep

Most solid modeling systems have adopted either the CSG or B-rep technique for representations of solids. Their advantages and disadvantages are summarized in Table 3.1 (Chiyokura 1988; McMahon and Jimmie 1998).

Table 3.1   Advantage and Disadvantages of CSG and B-Rep

Representation Type	CSG	B-rep
Advantages	The data file of CSG is concise. CSG guarantees, automatically, that objects created by CSG are valid. CSG is more user-friendly. Algorithms for converting CSG into B-rep have been developed.	The information is complete, especially for adjacent topology relations. Considerable number of methodologies developed for engineering analyses is based on B-rep. B-rep has the most refined geometric information compared to the others.
Disadvantages	The CSG database contains information about a solid in an unevaluated form. The validity of a feature of an object cannot be assessed without evaluating the entire tree. The tree is not unique for the same part design.	B-rep needs further feature extraction procedures to extract features from its face-edge-vertex database. The B-rep model does not provide any explicit information spatial constraints between features. The data structure of B-rep is complex compared with CSG.

3.12 Feature Recognition

Feature recognition involves the identification and grouping of feature entities from a geometric model. Such "post definition" of features can be done interactively or automatically. Usually, identified entities (i.e., the recognized features) are extracted from the model, and additional engineering information such as tolerances and nongeometric attributes are then associated with the feature entities. Figure 3.16 shows the steps of feature recognition.

Design by features is called feature-based design (FBD). This method uses a library of 2D or 3D features as design primitives at the product modeling level. The use of features provides a more natural interface between design and solid model (Wen and Ronk 1997). In mechanical design, for example, a designer can work directly with high-level entities such as a pocket rather than associating low-level entities in which vertices and edges form a pocket. Features allow the capability of providing additional information useful for process planning. Since features reflect specific manufacturing processes, they ensure that parts can be produced. Figure 3.17 shows the steps for design by feature procedure.

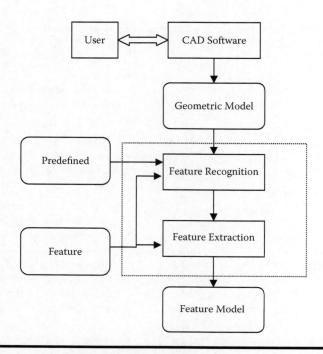

Figure 3.16   Feature recognition procedure.

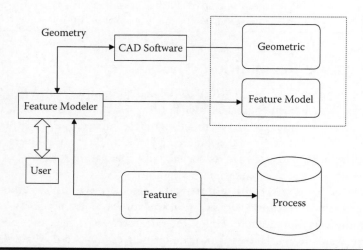

Figure 3.17   Feature-based design.

The integration of design by feature and feature recognition is applicable for design and manufacturing integration. A user can create a context-dependent representation that is specific and informative enough to satisfy the requirements of an application. While in the recognition process, the user builds the geometric model of a part that is successively mapped into the feature-based model (Martino et al. 1994).

Feature recognition is also used to directly recognize features from a CAD model of a part. It is easy to derive a geometric feature from the higher-level entities of a part model, where each feature is associated with a component of a solid model. On the other hand, to transform the geometric model into a feature representation is difficult. There are many methods that are used to extract features from a CAD model.

Interacting features are complex feature subparts on the part that cannot be recognized as any isolated standard feature. Shah and Mantyla (1995) defined feature interactions as *"feature interactions are intersections of feature boundaries with those of other features such that either the shape or the semantics of a feature are altered from the standard or generic definition."* An important distinction must be made between intersecting and interacting features. Interacting features are defined in the context of design by features, where design features are used to build the model. The addition (or subtraction) of a feature to the part can result in the creation of several new features due to interaction between the features. On the other hand, intersecting features are present on conventional representations and are not generated by feature operations (Wen and Ronak 1997). Algorithms are developed to extract both intersecting and interacting features. The feature interaction problem is usually treated as a problem of feature relationships. Although no existing feature recognition system is currently available to recognize 3D solid primitive features and their various interacting combinations (Allada and Anand 1996; Huikange et al. 2002).

3.13 Conclusion

In this chapter, feature definition, Wireframe Modeling, Surface Modeling, Boundary Representation (B-rep), Constructive Solid geometry (CGS), and definition of interacting features were discussed. Computer-Aided Design (CAD) is defined as any design activity that involves the effective use of the computer to create, to modify, to analyze, to optimize, and to document an engineering design. CAD is most commonly associated with the use of an interactive computer graphics system, referred to as a CAD system. The main advantages of using a CAD system are

■ *To increase the productivity of the designer*: To help the designer to conceptualize the product and its components. Therefore, this helps reduce the time required by the designer to create, to analyze, and to document the design.

■ *To improve the quality of the design*: Use of a CAD system with appropriate hardware and software capabilities permits the designer to perform a comprehensive engineering analysis and to involve a larger number and variety of design alternatives. The quality of the resulting design is thereby enhanced.

■ *To improve design documentation*: The final graphical output of a CAD system results in better documentation of the design than what is practical with manual drafting. The engineering drawings are superior, and there is more standardization among the drawings, fewer drafting errors, and greater legibility.

■ *To create a manufacturing database*: During the process of documentation the product design (geometric specification of the product, dimensions of the components, materials specifications, bill of materials, etc.), much of the required data for the manufacture the product is also created.

Author

Ali K. Kamrani is an Associate Professor of Industrial Engineering. He is also Founding Director of the Design and Free Form Fabrication Laboratory at the University of Houston, USA. He received his BS in Electrical Engineering in 1984, Master of Engineering in Electrical Engineering in 1985, Master of Engineering in Computer Science and Engineering Mathematics in 1987, and PhD in Industrial Engineering in 1991, all from the University of Louisville, Louisville, Kentucky. His research has been motivated by the fundamental application of systems engineering and its application in advanced design and development of complex systems. He is the Editor-in-Chief of the *International Journal of Collaborative Enterprise* and the *International Journal of Rapid Manufacturing*.

References

Allada, V., and S. Anand. 1996. Machine understanding of manufacturing features. *International Journal of Production Research* 34: 1791–1820.

Case, K., and W.A.W. Harun. 2000. Feature-based representation for manufacturing planning. *International Journal of Production Research* 38: 4285–4300.

Chandra, R.D., and K. Ghosh. 1999. Feature-based modeling and neural networks-based CAPP for integrated manufacturing. *International Journal of Computer Integrated Manufacturing* 12: 61–74.

Chang, T.C. 1990. *Expert Process Planning for Manufacturing*. Massachusetts: Addison-Wesley Publishing Company, Reading.

Chang, T.C., and R.A. Wysk. 1984. Integrated CAD and CAM through automated process planning. *International Journal of Production Research*: 158–172.

Chang, T.E., R.A. Wysk, and Wang, H.P. 1998. *Computer Aided Manufacturing*. New Jersey: Prentice Hall.

Chasen, S.H. 1986. Principles of geometric modeling. *CIM Technology*: 15–18.

Chiyokura, H. 1988. *Solid Modeling with Designbase: Theory and Implementation*. Reading, MA: Addison-Wesley Publishing Company.

Ciurana, J., and M.L.G. Romeu. 2003. Optimizing process planning using groups of precedence between operations based on machined volume. *Engineering Computations* 20: 67–81.

Devireddy, C.R., and K. Ghosh. 1999. Feature-based modeling and neural networks-based CAPP for integrated manufacturing. *International Journal of Computer Integrated Manufacturing* 12: 61–74.

Dixon, J., and J. Cunningham. 1988. Design with features: The origin of features. *Proceedings of the ASME Computer in Engineering Conference*. San Francisco, CA.

Groover, M.P. 2008. *Automation, Production Systems, and Computer-Integrated Manufacturing*. New Jersey: Prentice-Hall.

Huikange, K., M. Nandakumar, and J. Shah. 2002. CAD/CAM integration using machining features. *International Journal of Computer Integrated Manufacturing* 15: 296–318.

Kang, T.S., and B.O. Nnaji. 1993. Feature representation and classification for automatic process planning. *Journal of Manufacturing System* 12: 133–145.

Kobbelt, L.P., M. Botsch, U. Schwanecke, and H. Seidel. 2001. Feature sensitive surface extraction from volume data. *Proceedings of the ACM Siggraph Conference on Computer Graphics*: 57–66.

Luby, S.C., J.R. Dixon, and M.K. Simmons. 1986. Creating and using a feature database. *Computers in Mechanical Engineering* 5: 25–33.

Mäntylä, M. 1988. *Introduction to Solid Modeling*. Rockville, MI: Computer Science Press.

Mäntylä, M., N. Dana, and S. Jami. 1996. Challenges in features-based manufacturing research. *Communications of the ACM* 39: 77–85.

Martino, T.D., B. Falcidieno, F. Giannini, S. Hassinger, and J. Ovtcharova, 1994. Feature based modeling by integrated design and recognitions approaches. *Computer Aided Design* 26: 646–653.

McMahon, C., and B. Jimmie. 1998. *CAD/CAM: Principles, Practice and Manufacturing Management*. Reading, MA: Addison-Wesley.

Mortensen, M.E. 1985. *Geometric Modeling*. New York: John Wiley & Sons.

Natekar, D., X. Zhang, and G. Subbarayan. 2004. Constructive solid analysis: A hierarchal, geometry-based meshless analysis procedure for integrated design and analysis. *Computer Aided Design* 36: 473–486.

Pratt, M.J., and P.R. Wilson. 1985. Requirements for Support of form Features in a Solid Modeling System. Final Report, CAM-I Report R-85-ASPP.-01.

Requicha, A.A.G. 1980. Representations of rigid solids: Theory, methods, and systems. *Computing Survey* 12: 437–464.

Requicha, A.A.G., and S.G. Chan. 1986. Representation of geometric features, tolerances, and attributes in solid modelers based on constructive geometry. *IEEE Journal of Robotics and Automation* RA-2: 156–166.

Reqicha, A., and H. Voelcker. 1982. Solid modeling: A historical summary and contemporary assessment. *IEEE Computer Graphics and Applications* 2: 9–24.

Santosh, K., K. Shanker, and G.K. Lal. 2003. A generative process planning system for cold extrusion. *International Journal of Production Research* 41: 269–296.

Shah, J.J., and M. Mantyla. 1995. *Parametric and Feature-Based CAD/CAM-Concepts, Techniques, and Applications*. New York: John Wiley & Sons.

Sheu, L.C., and J.T. Lin. 1993. Representation scheme for defining and operating from features. *Computer Aided Design* 25: 333–347.

Somashekar, S., and W. Michael. 1995. An overview of automatic feature recognition techniques for computer-aided process planning. *Computers in Industry* 26: 1–21.

Sreevalsan, P.C., and J.J. Shah. 1992. Unification of form feature definition methods. In *Intelligent Computer Aided Design*, ed. D.C. Browns, M. Waldron, and H. Yoshikawa, 83–106. IFIP Transactions.

Taylor, D. 1992. *Computer Aided Design*. New York: Addison-Wesley Publishing Company.

Waco, D.L., and Y.C. Kim. 1994. Geometric reasoning for machining features using convex decomposition. *Computer Aided Design* 26: 477–489.

Wen, F.L., and M. Ronak. 1997. Feature-based design in an integrated CAD/CAM system for design for manufacturability of machining prismatic parts. *Concurrent Product Design and Environmentally Conscious Manufacturing* 5: 95–112.

Yong, Y., and R. Tang. 2000. Historical procedures and G-DSG method based manufacturing planning. *Chinese Journal of Aeronautics* 13: 123–128.

Zeid, I. 2007. *CAD/CAM Theory and Practice*. New York: McGraw-Hill.

Chapter 4

Selection of Parameters for CAD-VR Data Translation

Abdulaziz M. El-Tamimi,[1] Emad S. Abouel Nasr,[1,2] and Mustufa H. Abidi[1]

[1]Industrial Engineering Department, College of Engineering, King Saud University, Riyadh, Saudi Arabia

[2]Mechanical Engineering Department, Helwan University, Cairo, Egypt

Contents

Abstract

Virtual Reality (VR) technology is now mature enough to permit serious engineering applications. The combination of this new technology with CAD/CAM software systems will enhance the field of computer-aided design and engineering. The selection of parameters during the conversion of Computer-Aided Design (CAD) data to VR is an important step for achieving a high-quality virtual environment. During the process of data transfer, there are many parameters involved that need to be selected efficiently for better results. Therefore, to select the appropriate set of parameters during data conversion, Design of Experiments (DOE) techniques were applied. Based on the statistical analysis of the selected conversion parameters, a set of guidelines has been developed for parameter selection during the conversion process. Mathematical predictive models for selected responses have also been developed using the Response Surface Methodology.

4.1 Introduction

In the last decade, Virtual Reality (VR) has been considered as a young technology, but now it is mature enough to give some serious engineering applications. VR is a technology that is a combination of powerful digital computers with special hardware and software to simulate an alternate world or environment. The objective of VR is to simulate reality is a way that is as convincing as reality to the user and in such a way that he can interact with it (Liu et al. 2008).

For the application of VR in the field of engineering, mainly in manufacturing, the design data is in CAD format. Integration is perhaps one of the most complex issues in the field of VR systems. The CAD model is converted into VR, either directly, or through the in-between stage of a rendering package. Bourdakis explained that there is a trade-off between the amount of time spent reorganizing the CAD model to match the translator and the time spent optimizing the resultant VR model. "It is normal to spend a few hours or even days, hand-optimizing the translated file" (Bourdakis 1996). In the process of CAD to VR translation, the NURBS surfaces in CAD have to be converted to the polygonal model in VR. Therefore, a tessellation process is required that will result in the triangular model from the NURB surfaces. Parameters associated with the tessellation affect the resulting polygonal model. This tessellation process can be accomplished in several ways using various algorithms and software.

Complex and extremely detailed CAD data translate into excessively large VR models, but the computational time required to run these must not slow user movement to an undesirable level. Optimization to permit real-time viewing is accomplished by decreasing the information to be processed and hence diminishing the computational effort required during each simulation loop. Three different methods are recognized for CAD-VR data transfer: Library-Based approach, Database approach, and Straightforward translation approach (Whyte 2000). A library-based approach is to build up a library of standard parts or components that would be reused in the VR environment. It reduces time in repetitive work, but a lot of effort is required to build up the library. Both CAD and VR environments uses a central database to control component information in the database approach. The straightforward approach is a common way of transferring CAD to VR. The complete CAD model is translated to VR directly or through some intermediate stage. Therefore, a special exchange interface is required in this process.

4.2 Literature Background

The most explored research field on translating the information of CAD models into a VR model is using the data translation interface. Some commercial software (such as Deep Exploration, ProductView) and CAD systems (CATIA, Pro/E, etc.) provide the translation interface for graphics systems. Using the Virtual Reality Markup Language (VRML) file as a transformational file is very popular in research, as it is a developing standard for describing interactive three-dimensional scenes delivered across the World Wide Web (Wan et al. 2006). Antonishek et al. used VRML as a bridge between the CAD system and Virtual Workbench (Antonishek et al. 1998). Whyte et al. explored the technical issues related to the integrated use of CAD and virtual environments within the house-building construction industry and investigated the practical use of VRML (Whyte 2000). STEP, IGES, and so on, are graphical data exchange standards, forming the second transformational file, which are employed to translate complex CAD data. Lee presented a system focusing on shape representation and interoperability of product models for distributed virtual prototyping, where STEP was used as a means of transferring and sharing product models (Lee 2001). Otherwise, many other formats are also used to transfer the information of CAD models into other VR systems, such as OpenFlight, DXF, 3DS, SLP, and so on. Weyrich and Drew presented an approach of a "virtual workbench" for virtual assembly. The system used the professional modeling tool Multigen Creator, and the OpenFlight format as its data interface with the virtual environment (Weyrich and Drews 1999). Bourdot et al. presented an approach to the integration of VR and CAD, by developing a VR-CAD framework with multimodal immersive interaction (Bourdot et al. 2010).

From the literature survey, it can be concluded that most of the research is carried out on the integration issue of CAD and VR systems. Still, there is a need to investigate and explore the parameters involved in the translation process of CAD models to VR models. This chapter focuses on this issue and develops a set of guidelines for selecting the parameters during the CAD to VR conversion process.

4.3 Methodology Applied for Parameter Selection during CAD-VR Data Conversion

The Design of Experiments (DOE) techniques are applied for the proper selection of parameters during the conversion process. Figure 4.1 shows the block diagram of the methodology employed for the selection of the translation parameters.

DOE techniques are well-known and efficient techniques for selecting or optimizing the parameters in a process that affect the output response significantly. There are seven basic steps for applying the DOE to any process or experiment: Recognition of problem or problem statement; Choice of factors, levels, and ranges; Selection of response variables; Choice of experimental design; Performing the experiments; Statistical analysis of the data; and Conclusions and recommendations (Montgomery 2005). These seven steps have been applied to the CAD-VR data translation process.

4.3.1 Recognition of Problem

A valve cam assembly is selected for performing the experiments because its parts contain various geometry features such as circles, rectangles, and so on. Figure 4.2 shows the CAD model of the valve cam assembly.

In this chapter, a straightforward translation approach is adopted to convert the CAD model into the VR model. ProductView® adapters are selected as the intermediate interface to convert the CAD model from PTC Pro-e® to the PTC Division Mockup® format. A lot of parameters are associated with ProductView® adapters, which need to be selected competently for better-quality results. Because of the inappropriate selection of conversion parameters, a problem arises in the converted VR model: the circular edges of the model do not look smooth, or sometimes the complete geometry looks distorted. Figure 4.3 shows the problem in the converted VR model.

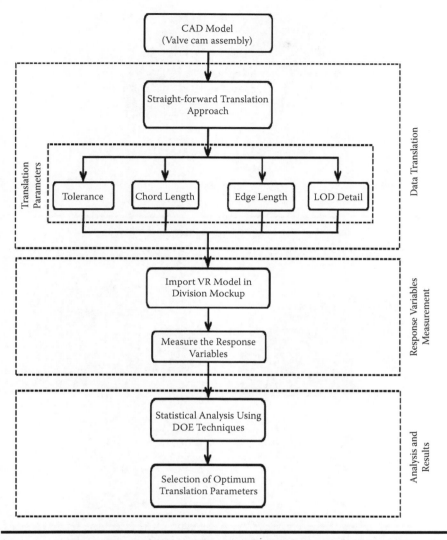

Figure 4.1   Block diagram representing the methodology employed for the selection of the translation parameters.

4.3.2 Choice of Factors, Levels, and Ranges

The selection of parameters and their levels is very important to study the performance characteristics of any product or process. So, it is a vital step to decide the factors and their levels for the experiment. It requires a thorough study of the system and experience on the system. Several preliminary experiments have been

Figure 4.2   CAD model of the selected valve cam assembly.

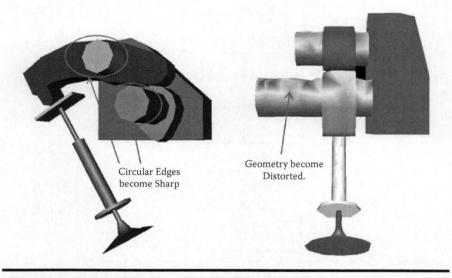

Circular Edges
become Sharp

Geometry become
Distorted.

Figure 4.3   Converted VR model of valve cam assembly with problems.

performed to select the most important factors. On the basis of preliminary experiments and experience, four parameters are identified, which are given in Table 4.1.

4.3.2.1 Tolerance

The tolerance specifies the maximum amount by which the tessellation is allowed to stray from the analytical surface (error or deviation).

Table 4.1 Parameters with Their Levels and Values

Factors	Number of Levels	Values
Tolerance	2	0.001 and 0.0007 (mm)
Chord Length	3	0.1, 0.3, and 0.5
Edge Length	3	0.1, 0.2, and 0.3
Number of Level of Detail (LOD)	3	3, 5, and 7

4.3.2.2 Chord Length

Also known as "chordal tolerance," this setting specifies the maximum distance between the surface of the original design and the tessellated surface of the STL triangle (the chord).

4.3.2.3 Edge Length

The edge length tolerance can be used to limit the maximum size of the output triangles. If the value of the edge length is smaller, then there will be more triangles in the polygonal model.

4.3.2.4 Number of Level of Detail (LOD)

The concept of LOD is that when a viewer is far away from the object, not all details have to be drawn.

4.3.3 Selection of Response Variables

Selection of the correct response variables in an experiment is crucial since it is a measure of the performance of the process under study. Response variables selected for the CAD-VR conversion process are given in Table 4.2.

4.3.3.1 Graphics Quality

It is the quality of the output model in the VR environment, that is, how it looks. It is the most important response in this process.

4.3.3.2 File Size

The space required by the converted model in the computer memory is the file size. It is recorded in kilobytes (KB).

Table 4.2 Response Variables for CAD-VR Conversion Process

Response	Unit
Graphics Quality	Very Good (1), Good (2) and Poor (3)
File Size	KB
Rendering Time	Seconds
Number of Triangles	Numbers

4.3.3.3 Rendering Time

The time taken by the computer to render the complete model is the rendering time.

4.3.3.4 Number of Triangles

The total number of triangles a converted polygonal model contains is known as the number of triangles.

4.3.4 Choice of Experimental Design and Analysis

Full factorial design analysis has been selected as the experimental design technique. With the given number of factors and their levels, full factorial design analysis is possible and the experiments it is possible to conduct with each and every combination of factors. On the other hand, Response Surface Methodology (RSM) is used to analyze the results more deeply, to build the predictive model, and to find the optimum value of parameters for a given set of response variables.

4.3.5 Performing the Experiments

4.3.5.1 Apparatus

The experiments were performed using a CAD model of a Valve Cam assembly. The computer used for running the experiments had the following specifications: Intel Core 2 Quad, Q9400 2.66 Hz 2.67 Hz processor with 4 GB RAM and NVIDIA GeForce 9800GT graphics card with 1 GB memory.

4.3.5.2 Procedure

The Division Mockup® software is used to measure the response variables. Table 4.3 shows the full factorial design table for the selected factors.

Table 4.3 Full Factorial Design Table with Responses

Run Order	Factors					Responses		
	Tolerance	Chord Length	Edge Length	LOD Detail	Graphics Quality	File Size (KB)	Rendering Time (s)	No. of Triangles
1	0.001	0.1	0.1	3	1	280	0.40	6544
2	0.001	0.1	0.1	5	2	272	0.34	5772
3	0.001	0.1	0.1	7	2	264	0.31	5568
4	0.001	0.1	0.2	3	1	228	0.23	6030
5	0.001	0.1	0.2	5	2	248	0.35	5244
6	0.001	0.1	0.2	7	2	252	0.43	4994
7	0.001	0.1	0.3	3	2	212	0.31	5342
8	0.001	0.1	0.3	5	3	210	0.28	4676
9	0.001	0.1	0.3	7	3	208	0.28	4450
10	0.001	0.3	0.1	3	2	268	0.35	5432
11	0.001	0.3	0.1	5	2	266	0.27	5324
12	0.001	0.3	0.1	7	2	260	0.41	5284
13	0.001	0.3	0.2	3	2	240	0.29	4976
14	0.001	0.3	0.2	5	2	236	0.23	4798

(Continued)

Table 4.3 (*Continued*) Full Factorial Design Table With Responses

	Factors					Responses			
Run Order	Tolerance	Chord Length	Edge Length	LOD Detail	Graphics Quality	File Size (KB)	Rendering Time (s)	No. of Triangles
15	0.001	0.3	0.2	7	2	256	0.30	4852
16	0.001	0.3	0.3	3	3	208	0.18	3708
17	0.001	0.3	0.3	5	3	200	0.22	3502
18	0.001	0.3	0.3	7	3	212	0.25	3634
19	0.001	0.5	0.1	3	2	280	0.28	5306
20	0.001	0.5	0.1	5	2	276	0.35	5232
21	0.001	0.5	0.1	7	2	279	0.32	5262
22	0.001	0.5	0.2	3	2	272	0.38	4804
23	0.001	0.5	0.2	5	3	256	0.31	4798
24	0.001	0.5	0.2	7	3	264	0.28	4754
25	0.001	0.5	0.3	3	3	220	0.23	3606
26	0.001	0.5	0.3	5	3	224	0.29	2642
27	0.001	0.5	0.3	7	3	216	0.34	3592
28	0.0007	0.1	0.1	3	1	272	0.31	7328

29	0.0007	0.1	0.1	5	1	276	0.30	6310
30	0.0007	0.1	0.1	7	1	252	0.29	5986
31	0.0007	0.1	0.2	3	1	248	0.23	7090
32	0.0007	0.1	0.2	5	1	244	0.25	6012
33	0.0007	0.1	0.2	7	2	256	0.28	5582
34	0.0007	0.1	0.3	3	1	240	0.24	6738
35	0.0007	0.1	0.3	5	2	216	0.23	5662
36	0.0007	0.1	0.3	7	2	210	0.22	5070
37	0.0007	0.3	0.1	3	2	272	0.24	5804
38	0.0007	0.3	0.1	5	2	260	0.26	5484
39	0.0007	0.3	0.1	7	2	264	0.32	5468
40	0.0007	0.3	0.2	3	2	252	0.29	5448
41	0.0007	0.3	0.2	5	2	256	0.20	5184
42	0.0007	0.3	0.2	7	2	260	0.35	5212
43	0.0007	0.3	0.3	3	2	236	0.25	4888
44	0.0007	0.3	0.3	5	3	244	0.30	4686
45	0.0007	0.3	0.3	7	3	240	0.29	4568

(Continued)

Table 4.3 (*Continued*) Full Factorial Design Table with Responses

	Factors					Responses			
Run Order	Tolerance	Chord Length	Edge Length	LOD Detail	Graphics Quality	File Size (KB)	Rendering Time (s)	No. of Triangles
46	0.0007	0.5	0.1	3	2	296	0.20	5548
47	0.0007	0.5	0.1	5	2	288	0.33	5452
48	0.0007	0.5	0.1	7	2	292	0.34	5458
49	0.0007	0.5	0.2	3	2	280	0.32	5170
50	0.0007	0.5	0.2	5	2	284	0.34	5160
51	0.0007	0.5	0.2	7	2	280	0.29	5150
52	0.0007	0.5	0.3	3	3	268	0.27	4700
53	0.0007	0.5	0.3	5	3	264	0.21	4698
54	0.0007	0.5	0.3	7	3	260	0.36	4612

4.3.6 Statistical Analysis of Results

4.3.6.1 Full Factorial Analysis

In order to assess the influence of selected factors on responses, full factorial analysis is done using the MINITAB 16 software. Using the full factorial analysis, which factor has a strong effect on which response and how the factors' interactions affect the responses will be investigated. For Graphics Quality, Figure 4.4 shows the main effects plot of selected factors on the graphics quality.

Based on the above figure, the following set of parameters is recommended: tolerance 0.0007, chord length 0.1, edge length 0.1, and number of LOD detail 3, for achieving better graphics quality. Based on a slope analysis of the main effects plot, it can be concluded that edge length is the most significant factor for graphics quality. Further conclusions will be drawn from ANOVA test (p-value); see Table 4.4.

After performing the ANOVA analysis (with $\alpha = 0.05$), it can be concluded that all the factors are significant (i.e., $p < \alpha$). Second-order interactions that are significant are between chord length and LOD Detail. All the third-order interactions are insignificant except interaction between tolerance, chord length, and edge length. For file size, Figure 4.5 shows the main effects plot of selected factors on the file size.

Based on the above figure, the following set of parameters is recommended for reducing the file size: tolerance 0.001, chord length 0.1, edge length 0.3, and number of LOD details 5. Based on slope analysis of the main effects plot, it can also be concluded that edge length is the most significant factor for file size. The number of

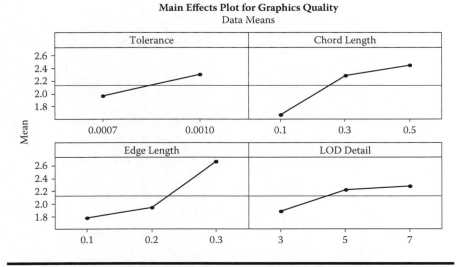

Figure 4.4 Main effects plot of selected factors for graphics quality.

Table 4.4 Analysis of Variance for Graphics Quality

Source	DF	Seq SS	Adj SS	Adj MS	F	P
Tolerance	1	1.500000	1.500000	1.500000	36.00	**0.000**
Chord Length	2	6.037040	6.037040	3.018520	72.44	**0.000**
Edge Length	2	8.037040	8.037040	4.018520	96.44	**0.000**
LOD Detail	2	1.592590	1.592590	0.796300	19.11	**0.001**
Tolerance*Chord Length	2	0.777780	0.777780	0.388890	9.33	**0.008**
Tolerance*Edge Length	2	0.111110	0.111110	0.055560	1.33	0.316
Tolerance*LOD Detail	2	0.111110	0.111110	0.055560	1.33	0.316
Chord Length*Edge Length	4	0.185190	0.185190	0.046300	1.11	0.415
Chord Length*LOD Detail	4	0.962960	0.962960	0.240740	5.78	**0.017**
Edge Length*LOD Detail	4	0.296300	0.296300	0.074070	1.78	0.226
Tolerance*Chord Length*Edge Length	4	0.777780	0.777780	0.194440	4.67	**0.031**
Tolerance*Chord Length*LOD Detail	4	0.444440	0.444440	0.111110	2.67	0.111
Tolerance*Edge Length*LOD Detail	4	0.444440	0.444440	0.111110	2.67	0.111
Chord Length*Edge Length*LOD Detail	8	0.481480	0.481480	0.060190	1.44	0.308
Error	8	0.333330	0.333330	0.041670		
Total	53	22.092590				
S = 0.204124 R-Sq = 98.49% R-Sq (adj) = 90.00%						

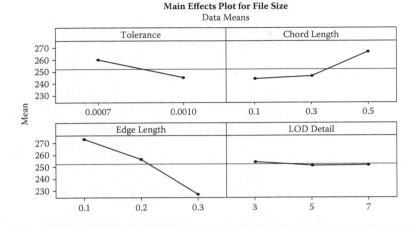

Figure 4.5   Main effects plot of selected factors for file size.

LOD details has very little effect on the file size. Further conclusions will be drawn from the ANOVA test (p-value); see Table 4.5.

After performing the ANOVA analysis, it can be concluded that all the factors are significant except LOD Detail. Second-order interactions that are significant are between tolerance and chord length; and tolerance and edge length. For rendering time, Figure 4.6 shows the main effects plot of selected factors on the rendering time.

Based on the above figure, the following set of parameters is recommended for reducing rendering time: tolerance 0.0007, chord length 0.3, edge length 0.3, and number of LOD details 3. Based on slope analysis of the main effects plot, it can also be concluded that edge length is the most significant factor for rendering time. Chord length has less effect on rendering time. Further conclusions will be drawn from the ANOVA test (p-value); see Table 4.6.

After performing the ANOVA analysis (with $\alpha = 0.05$), it can be concluded that only some of the factors are significant, such as tolerance, edge length, and LOD detail. All the second-order interactions are insignificant. All the third-order interactions are insignificant except chord length, edge length, and LOD detail interaction. For number of triangles, Figure 4.7 shows the main effects plot of selected factors on the number of triangles.

Based on above figure, it can be seen that for reducing the number of triangles, the following set of parameters is recommended: tolerance 0.001, chord length 0.5, edge length 0.3, and number of LOD details 7. Based on slope analysis of the main effects plot, it can also be concluded that edge length is the most significant factor for the number of triangles, and the number of LOD details has less effect on the number of triangles. Further conclusions will be drawn from the ANOVA test (p-value); see Table 4.7.

Table 4.5   Analysis of Variance for File Size

Source	DF	Seq SS	Adj SS	Adj MS	F	P
Tolerance	1	3007.57	3007.57	3007.57	53.40	**0.000**
Chord Length	2	5682.33	5682.33	2841.17	50.44	**0.000**
Edge Length	2	19534.11	19534.11	9767.06	173.41	**0.000**
LOD Detail	2	91.44	91.44	45.72	0.81	0.478
Tolerance*Chord Length	2	951.81	951.81	475.91	8.45	**0.011**
Tolerance*Edge Length	2	1671.15	1671.15	835.57	14.84	**0.002**
Tolerance*LOD Detail	2	85.81	85.81	42.91	0.76	0.498
Chord Length*Edge Length	4	308.22	308.22	77.06	1.37	0.326
Chord Length*LOD Detail	4	173.56	173.56	43.39	0.77	0.574
Edge Length*LOD Detail	4	541.78	541.78	135.44	2.40	0.135
Tolerance*Chord Length*Edge Length	4	212.96	212.96	53.24	0.95	0.486
Tolerance*Chord Length*LOD Detail	4	151.63	151.63	37.91	0.67	0.629
Tolerance*Edge Length*LOD Detail	4	30.96	30.96	7.74	0.14	0.964
Chord Length*Edge Length*LOD Detail	8	565.56	565.56	70.69	1.26	0.378
Error	8	450.59	450.59	56.32		
Total	53	33459.5				
S = 7.50494 R-Sq = 98.65% R-Sq (adj) = 91.08%						

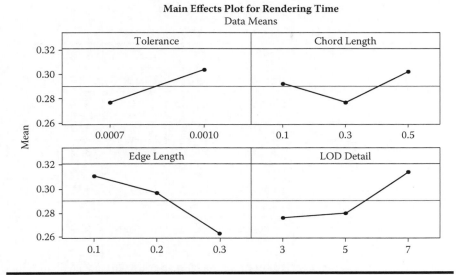

Figure 4.6   Main effects plot of selected factors for rendering time.

After performing the ANOVA analysis, it can be concluded that all the factors and their second-order interactions are significant except interaction between tolerance and chord length and edge length and LOD detail. All the third-order interactions are insignificant except the interaction between tolerance, chord length, and edge length.

4.3.6.2 *Response Surface Methodology (RSM)*

RSM is a collection of mathematical and statistical techniques that are useful for the modeling and analysis of problems in which a response of interest is influenced by several variables and the objective is to optimize this response (Montgomery 2005). RSM is used in this work to generate the predictive models for the various responses. In this section, the factors are denoted as: Tolerance = T, Chord Length = CL, Edge Length = EL, and No. of LOD Details = LD.

For Graphics Quality: From the RSM analysis, a reduced predictive model for graphics quality is given in Equation 4.1:

$$\begin{aligned}
Graphics\ Quality = {} & -1.76852 + 2222.22 * T + 10.5093 * CL - 6.66667 * EL \\
& + 0.222222 * LD - 5.55556 * CL^2 + 27.7778 * EL^2 - 3703.7 \\
& * T * CL - 0.416667 * CL * LD \tag{4.1}
\end{aligned}$$

Figure 4.8 shows the response surface of graphics quality.

Table 4.6   Analysis of Variance for Rendering Time

Source	DF	Seq SS	Adj SS	Adj MS	F	P
Tolerance	1	0.009779	0.009779	0.009779	5.63	**0.045**
Chord Length	2	0.005814	0.005814	0.002907	1.67	0.247
Edge Length	2	0.021882	0.021882	0.010941	6.30	**0.023**
LOD Detail	2	0.015704	0.015704	0.007852	4.52	**0.049**
Tolerance*Chord Length	2	0.011212	0.011212	0.005606	3.23	0.094
Tolerance*Edge Length	2	0.005200	0.005200	0.002600	1.50	0.280
Tolerance*LOD Detail	2	0.000425	0.000425	0.000213	0.12	0.886
Chord Length*Edge Length	4	0.006001	0.006001	0.001500	0.86	0.525
Chord Length*LOD Detail	4	0.007934	0.007934	0.001984	1.14	0.403
Edge Length*LOD Detail	4	0.001240	0.001240	0.000310	0.18	0.943
Tolerance*Chord Length*Edge Length	4	0.010161	0.010161	0.002540	1.46	0.299
Tolerance*Chord Length*LOD Detail	4	0.002529	0.002529	0.000632	0.36	0.828
Tolerance*Edge Length*LOD Detail	4	0.004922	0.004922	0.001230	0.71	0.608
Chord Length*Edge Length*LOD Detail	8	0.050566	0.050566	0.006321	3.64	**0.043**
Error	8	0.013894	0.013894	0.001737		
Total	53	0.167263				
S = 0.0416747 R-Sq = 91.69% R-Sq (adj) = 44.97%						

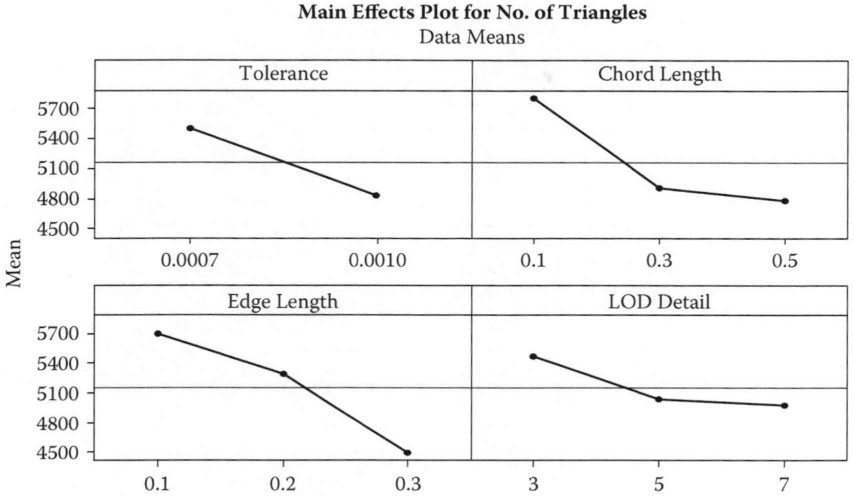

Main Effects Plot for No. of Triangles
Data Means

Figure 4.7   Main effects plot of selected factors for number of triangles.

From Figure 4.8, it can be seen that, for better graphics quality, the factors should be as follows: T = 0.0007, CL = 0.1, EL = 0.1, and LD = 3. Table 4.8 shows the ANOVA table for a reduced model with a 5% level of significance (α = 0.05) for graphics quality analysis.

From ANOVA, it can be concluded that all the linear terms are significant, while some of the square terms and interaction terms are also significant. The insignificant terms are left out from the model. The value of R-square suggests that the mathematical model for graphics quality has a sufficient predictive power, and it explains 78.43% variation in the data.

For File Size: Based on the RSM, a reduced predictive model for file size is given in Equation 4.2:

$$File\ Size = 197.941 + 90895.1 * T + 66.4352 * CL + 392.407 * EL + 227.083$$
$$* CL^2 - 608.333 * EL^2 - 171296 * T * CL - 446296 * T * EL \quad (4.2)$$

Figure 4.9 shows the response surface of file size.

From Figure 4.9, it can be seen that, for lesser file size, the factors should be as follows: T = 0.0007, CL = 0.1, and EL = 0.1. Table 4.9 shows the ANOVA table for the reduced model with 5% level of significance (α = 0.05) for file size analysis.

From the initial ANOVA, it is found that LOD detail and its interaction terms are insignificant; therefore, it is left out from the reduced model. From the ANOVA of the reduced model, it can be concluded that all the linear terms, square terms, and interaction terms are significant. The insignificant terms are left out from the model. Lack of fit is also insignificant. The high value of R-square suggests that

Table 4.7   Analysis of Variance for Number of Triangles

Source	DF	Seq SS	Adj SS	Adj MS	F	P
Tolerance	1	6230166	6230166	6230166	376.61	**0.000**
Chord Length	2	11232783	11232783	5616392	339.51	**0.000**
Edge Length	2	13663919	13663919	6831959	412.99	**0.000**
LOD Detail	2	2646945	2646945	1323473	80	**0.000**
Tolerance*Chord Length	2	105228	105228	52614	3.18	0.096
Tolerance*Edge Length	2	1656304	1656304	828152	50.06	**0.000**
Tolerance*LOD Detail	2	165468	165468	82734	5	**0.039**
Chord Length*Edge Length	4	531393	531393	132848	8.03	**0.007**
Chord Length*LOD Detail	4	2553697	2553697	638424	38.59	**0.000**
Edge Length*LOD Detail	4	38422	38422	9605	0.58	0.685
Tolerance*Chord Length*Edge Length	4	314116	314116	78529	4.75	**0.029**
Tolerance*Chord Length*LOD Detail	4	186320	186320	46580	2.82	0.099
Tolerance*Edge Length*LOD Detail	4	130459	130459	32615	1.97	0.192
Chord Length*Edge Length*LOD Detail	8	151364	151364	18921	1.14	0.427
Error	8	132341	132341	16543		
Total	53	39738925				
S = 128.618 R-Sq = 99.67% R-Sq (adj) = 97.79%						

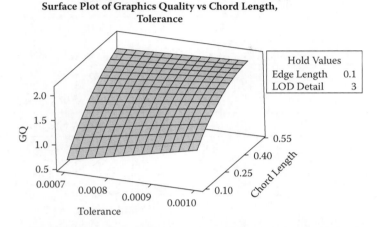

Figure 4.8   Surface plot of graphics quality.

the mathematical model for file size has a sufficient predictive power, and it explains only 90.8% variation in the data.

For Rendering Time: From the RSM analysis, a reduced predictive model for rendering time is given in Equation 4.3:

$$Rendering\ Time = 0.214501 + 89.7119 * T - 0.239815 * EL$$
$$+ 0.00949074 * LD \tag{4.3}$$

Figure 4.10 shows the response surface of rendering time.

Based on Figure 4.10, it can be seen that for lesser rendering time, the factors should be as follows: T = 0.0007, EL = 0.1, and LOD = 3. Table 4.10 shows the ANOVA table for the reduced model with 5% level of significance ($\alpha = 0.05$) for rendering time analysis.

Based on the initial ANOVA, it is found that chord length, its interaction terms, and all square terms are insignificant; therefore, they are left out from the reduced model. From the ANOVA of the reduced model, it can be concluded that all the available linear terms are significant. Lack of fit is also insignificant. The value of R-square suggests that the mathematical model for rendering time may not provide sufficient predictive power as it explains only 21.5% variation in the data.

For Number of Triangles: From the RSM analysis, a reduced predictive model for the number of triangles is given in Equation 4.4:

$$Number\ of\ Triangles = 9108.02 + 460000 * T - 10852.6 * CL + 15278.2$$
$$* EL - 811.208 * LD + 9609.72 * CL^2 - 19944.4 * EL^2$$
$$+ 46.4306 * LD^2 - 13622222 * T * EL - 5912.5 * CL$$
$$* EL + 741.25 * CL * LD \tag{4.4}$$

Table 4.8 Analysis of Variance for Graphics Quality

Source	DF	Seq SS	Adj SS	Adj MS	F	P
Regression	8	18.0463	18.0463	2.2558	25.0900	**0.0000**
Linear	4	15.4167	15.4167	3.8542	42.8600	**0.0000**
Tolerance	1	1.5000	1.5000	1.5000	16.6800	**0.0000**
Chord Length	1	5.4444	5.4444	5.4444	60.5500	**0.0000**
Edge Length	1	7.1111	7.1111	7.1111	79.0800	**0.0000**
LOD Detail	1	1.3611	1.3611	1.3611	15.1400	**0.0000**
Square	2	1.5185	1.5185	0.7593	8.4400	**0.0010**
Chord Length*Chord Length	1	0.5926	0.5926	0.5926	6.5900	**0.0140**
Edge Length*Edge Length	1	0.9259	0.9259	0.9259	10.3000	**0.0020**
Interaction	2	1.1111	1.1111	0.5556	6.1800	**0.0040**
Tolerance*Chord Length	1	0.4444	0.4444	0.4444	4.9400	**0.0310**
Chord Length*LOD Detail	1	0.6667	0.6667	0.6667	7.4100	**0.0090**
Residual Error	45	4.0463	4.0463	0.0899		
Total	53	22.0926				
R-Sq = 81.68% R-Sq(pred) = 73.37% R-Sq(adj) = 78.43%						

Figure 4.11 shows the response surface of rendering time.

From Figure 4.11, it can be seen that for a lesser number of triangles, the factors should be as follows: T = 0.001, CL = 0.5, EL = 0.1, and LOD = 3. Table 4.11 shows the ANOVA table for the reduced model with a 5% level of significance ($\alpha = 0.05$) for the number of triangles analysis.

From the initial ANOVA, it is found that interactions of tolerance and chord length, tolerance and LOD detail, and edge length and LOD detail are insignificant. From the ANOVA of the reduced model, it can be concluded that all the available linear terms, square terms, and interaction terms are significant. The insignificant terms are left out from the model. The high value of R-square suggests that the mathematical model for the number of triangles has sufficient predictive power, and it explains only 93.75% variation in the data.

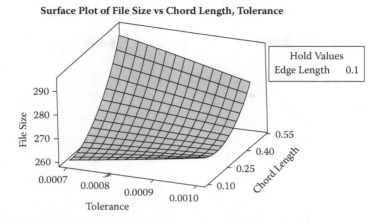

Figure 4.9   Surface plot of file size.

4.3.6.2.1 Analysis of Predictive Models

A paired t-test is conducted to compare the actual response values and predicted response values.

For Graphics Quality: To fulfill the key assumption behind the paired t-test, the Anderson–Darling test is done to check the normality of the data (i.e., difference between actual and predicted response values). The p-value is 0.245, that is, greater than $\alpha = 0.05$. Therefore, the hypothesis of normality cannot be rejected. Table 4.12 shows the paired t-test statistics.

From the paired t-test for graphics quality, it is concluded that there is no significant difference between the mean value of actual response and the predicted response. There is a difference in the standard deviation of the actual response and the predicted response. The results suggest that the mathematical model can predict accurate response values. A graphical comparison of the actual and predicted response of graphics quality is shown in Figure 4.12.

From the above figure, it can be seen that there is not a noteworthy difference between actual and predicted graphics quality for most of the data points.

For File Size: The p-value from the Anderson–Darling test for normality is 0.635, that is, greater than $\alpha = 0.05$. Therefore, the hypothesis of normality cannot be rejected. Table 4.13 shows the paired t-test statistics.

From the paired t-test statistics for file size, it is concluded that there is not a significant difference between the mean value of actual response and the predicted response. There is a difference in the standard deviation of actual response and the predicted response. The results suggest that the mathematical model can predict accurate response values. Figure 4.13 shows the graphical comparison of actual and predicted responses of file size. Based on Figure 4.13, it can be seen that almost all

Table 4.9 Analysis of Variance for File Size

Source	DF	Seq SS	Adj SS	Adj MS	F	P
Regression	7	30788.1000	30788.1000	4398.3000	75.7400	**0.0000**
Linear	3	26789.9000	26789.9000	8930.0000	153.7700	**0.0000**
Tolerance	1	3007.6000	3007.6000	3007.6000	51.7900	**0.0000**
Chord Length	1	4692.2000	4692.2000	4692.2000	80.8000	**0.0000**
Edge Length	1	19090.0000	19090.0000	19090.0000	328.7200	**0.0000**
Square	2	1434.2000	1434.2000	717.1000	12.3500	**0.0000**
Chord Length*Chord Length	1	990.1000	990.1000	990.1000	17.0500	**0.0000**
Edge Length*Edge Length	1	444.1000	444.1000	444.1000	7.6500	**0.0080**
Interaction	2	2564.1000	2564.1000	1282.0000	22.0800	**0.0000**
Tolerance*Chord Length	1	950.7000	950.7000	950.7000	16.3700	**0.0000**
Tolerance*Edge Length	1	1613.4000	1613.4000	1613.4000	27.7800	**0.0000**
Residual Error	46	2671.4000	2671.4000	58.1000		
Lack-of-Fit	10	580.1000	580.1000	58.0000	1.0000	0.4630
Pure Error	36	2091.3000	2091.3000	58.1000		
Total	53	33459.5000				

R-Sq = 92.02% R-Sq(pred) = 88.92% R-Sq(adj) = 90.80%

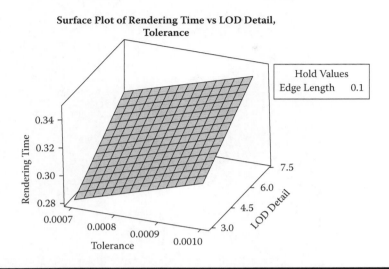

Figure 4.10   Surface plot of rendering time.

Table 4.10   Analysis of Variance for Rendering Time

Source	DF	Seq SS	Adj SS	Adj MS	F	P
Regression	3	0.0435	0.0435	0.0145	5.8500	**0.0020**
Linear	3	0.0435	0.0435	0.0145	5.8500	**0.0020**
Tolerance	1	0.0098	0.0098	0.0098	3.9500	**0.0520**
Edge Length	1	0.0207	0.0207	0.0207	8.3600	**0.0060**
LOD Detail	1	0.0130	0.0130	0.0130	5.2400	**0.0260**
Residual Error	50	0.1238	0.1238	0.0025		
Lack-of-Fit	14	0.0157	0.0157	0.0011	0.3700	0.9740
Pure Error	36	0.1081	0.1081	0.0030		
Total	53	0.1673				
R-Sq = 25.98% R-Sq(pred) = 13.70% R-Sq(adj) = 21.54%						

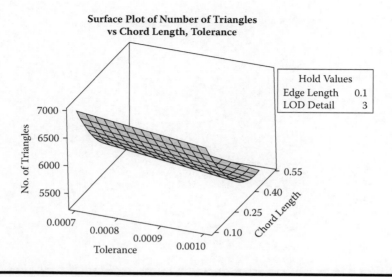

Figure 4.11   Surface plot of number of triangles.

the data points of actual and predicted file sizes are overlapping with one another; therefore, there is no sign of significant difference.

For Rendering Time: The p-value from the Anderson–Darling test for normality is 0.438, that is, greater than $\alpha = 0.05$. Therefore, the hypothesis of normality cannot be rejected. The paired t-test statistics is shown in Table 4.14.

From the paired t-test statistics for rendering time, it is concluded that there is not a significant difference between the mean value of actual response and the predicted response. There is a difference in the standard deviation of actual response and the predicted response. The results suggest that the mathematical model can predict accurate response values. Figure 4.14 shows the graphical comparison of actual and predicted response of rendering time.

From Figure 4.14, it can be seen that there is a difference between actual and predicted rendering time for most of the data points, but the difference is not very high.

For Number of Triangles: The p-value from the Anderson–Darling test for normality is 0.456, that is, greater than $\alpha = 0.05$. Therefore, the hypothesis of normality cannot be rejected. The paired t-test statistics is shown in Table 4.15.

From the paired t-test statistics for the number of triangles, it is concluded that there is not a significant difference between the mean value of actual response and the predicted response. There is a difference in the standard deviation of the actual response and the predicted response. The results suggest that the mathematical model can predict accurate response values. Figure 4.15 shows the graphical comparison of actual and predicted responses of the number of triangles.

Table 4.11   Analysis of Variance for Number of Triangles

Source	DF	Seq SS	Adj SS	Adj MS	F	P
Regression	10	37722377.00	37722377.00	3772238.00	80.4400	**0.0000**
Linear	4	31109505.00	31109505.00	7777376.00	165.8400	**0.0000**
Tolerance	1	6230166.00	6230166.00	6230166.00	132.8500	**0.0000**
Chord Length	1	9459725.00	9459725.00	9459725.00	201.7200	**0.0000**
Edge Length	1	13186582.00	13186582.00	13186582.00	281.1800	**0.0000**
LOD Detail	1	2233032.00	2233032.00	2233032.00	47.6200	**0.0000**
Square	3	2664308.00	2664308.00	888103.00	18.9400	**0.0000**
Chord Length*Chord Length	1	1773058.00	1773058.00	1773058.00	37.8100	**0.0000**
Edge Length*Edge Length	1	477337.00	477337.00	477337.00	10.1800	**0.0030**
LOD Detail*LOD Detail	1	413913.00	413913.00	413913.00	8.8300	**0.0050**
Interaction	3	3948563.00	3948563.00	1316188.00	28.0700	**0.0000**
Tolerance*Edge Length	1	1503076.00	1503076.00	1503076.00	32.0500	**0.0000**
Chord Length*Edge Length	1	335594.00	335594.00	335594.00	7.1600	**0.0110**
Chord Length*LOD Detail	1	2109894.00	2109894.00	2109894.00	44.9900	**0.0000**
Residual Error	43	2016548.00	2016548.00	46896.00		
Total	53	39738925.00				
R-Sq = 94.93% R-Sq(pred) = 91.90% R-Sq(adj) = 93.75%						

Table 4.12   Paired t-test Statistics for Graphics Quality

	N	Mean	StDev	SE Mean	
Actual Graphics Quality	54	2.1296	0.6456	0.0879	
Predicted Graphics Quality	54	2.1296	0.5835	0.0794	
Difference	54	0	0.2763	0.0376	
95% CI for mean difference: (−0.0754, 0.0754)					
t-test of mean difference = 0 (vs not = 0): *t*-value = −0.00 *p*-value = 1.000					

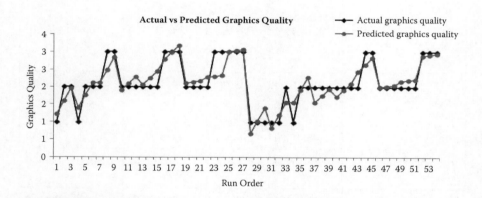

Figure 4.12   Graphical comparison of actual and predicted graphics quality.

Table 4.13   Paired t-test Statistics for File Size

	N	Mean	StDev	SE Mean	
Actual File Size	54	252.17	25.13	3.4200	
Predicted File Size	54	252.17	24.10	3.2800	
Difference	54	0	7.10	0.9660	
95% CI for mean difference: (−1.938, 1.938)					
t-test of mean difference = 0 (vs. not = 0): *t*-value = 0.00 *p*-value = 1.000					

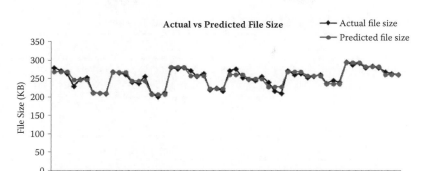

Figure 4.13   Graphical comparison of actual and predicted file size.

Table 4.14   Paired t-test Statistics for Rendering Time

	N	Mean	StDev	SE Mean	
Actual Rendering Time	54	0.29025	0.05618	0.0076	
Predicted Rendering Time	54	0.29025	0.02863	0.0039	
Difference	54	0	0.04833	0.0066	
95% CI for mean difference: (−0.01319, 0.01319)					
t-test of mean difference = 0 (vs. not = 0): t-value = 0.00 p-value = 1.000					

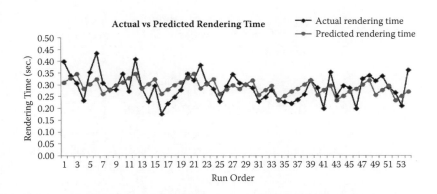

Figure 4.14   Graphical comparison of actual and predicted rendering time.

Table 4.15 Paired t-test Statistics for Number of Triangles

	N	Mean	StDev	SE Mean	
Actual No. of Triangles	54	5159	866	118	
Predicted No. of Triangles	54	5159	844	115	
Difference	54	0	195.1	26.5	
95% CI for mean difference: (−53.2, 53.2)					
t-test of mean difference = 0 (vs. not = 0): *t*-value = −0.00 *p*-value = 1.000					

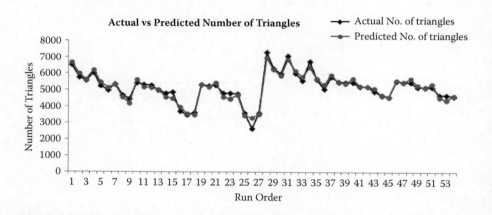

Figure 4.15 Graphical comparison of actual and predicted number of triangles.

From the above figure, it can be seen that almost all the data points of the actual and predicted number of triangles overlap each other; therefore, there is no indication of significant difference.

4.4 Conclusions and Recommendations

The effect of important CAD-VR translation parameters were investigated to determine the optimum values for these parameters in order to achieve the best graphics quality with minimum file size, rendering time, and number of triangles. The translation parameters that largely affect the graphics quality were tolerance, chord length, and edge length. The study also reveals that the number of LOD details has the least influence on responses except the rendering time. Based on

Figure 4.16   VR model of valve cam assembly after applying the recommended parameters.

the study, the following values are recommended for various CAD-VR translation parameters in order to achieve better graphics quality:

Tolerance = 0.0007, Chord Length = 0.1, Edge Length = 0.1 and Number of LOD Details = 3.

The above parameters results in a very good graphics quality, but the file size, number of triangles, and rendering time also increase. The CAD models that contain less geometrical complexity such as simple cubical parts and so on do not require a high number of triangles tessellated model, and still result in a better-graphics-quality VR model. Therefore, for the CAD models with simple features, the following set of parameters is recommended:

Tolerance = 0.001, Chord Length = 0.5, Edge Length = 0.3, and Number of LOD details = 7.

The predicted models for all the responses can predict the values to a satisfactory level of accuracy except the model of the rendering time. After applying the recommended parameters during the conversion process of the selected CAD model, that is, Valve cam assembly, the resulting VR model is shown in Figure 4.16.

Authors

Abdulaziz M. El-Tamimi is an emeritus professor of manufacturing systems engineering at the Department of Industrial Engineering, King Saud University. He has more than 43 years' experience in research, teaching, consultations, and training for the design and analysis of industrial and production systems. This involves various activities and tasks that include machine design, instrumentation and control, computer control, computer-aided design and manufacturing, knowledge-base design, facility design of production systems, maintenance, quality, and safety applications with strong background in organizing, planning and management of engineering and training programs, engineering projects, and technology transfer. He has vast experience in system analysis and design. He holds a PhD from the University of Manchester, Institute of Science and Technology, Manchester, U.K.

Emad S. Abouel Nasr is an Associate Professor in Industrial Engineering Department, College of Engineering, King Saud University, and Assistant Professor in Mechanical Engineering Department at Helwan University, Faculty of Engineering, Helwan, Cairo, Egypt. He received his PhD in Industrial Engineering from the University of Houston, Texas, in 2005. His current research focuses on CAD, CAM, rapid prototyping, advanced manufacturing systems, and collaborative engineering.

Mustufa H. Abidi is currently a researcher at the Advanced Manufacturing Institute at the Department of Industrial Engineering, King Saud University. He pursued his master's degree in Industrial Engineering from King Saud University. He graduated from Jamia Millia Islamia, New Delhi, India. He received a gold medal from the Faculty of Engineering and Technology, Jamia Millia Islamia. The application of virtual reality techniques for sustainable product development is the major focus of his research; other than that, he is actively involved in research areas that include Flexible Manufacturing Systems, Micro-Manufacturing, Human-Computer Interaction, and Reverse engineering. He has obtained Six-Sigma Green Belt Certificate and Certified Supply Chain Manager Certificate.

References

Antonishek, B., Egts, D.D., and U.R. Obeysekare 1998. Virtual assembly using two-handed interaction techniques on the virtual workbench. *Proceedings of the 1998 ASME Design Technical Conference and Computer in Engineering Conference* 13–16.

Bourdakis, V. 1996. From CAAD to VR building a VRML model of London's West End. *Proceedings of 3rd-UK VRSIG Conference.* De Montford University, Leicester.

Bourdot, P., Convard, T., Picon, F., Ammi, M., Touraine, D., and J.M. Vézien, 2010. VRCAD integration: Multimodal immersive interaction and advanced haptic paradigms for implicit edition of CAD models. *Computer-Aided Design* 42: 445–61.

Lee, J.Y. 2001. Shape representation and interoperability for virtual prototyping in a distributed design environment. *International Journal of Advanced Manufacturing Technology* 17: 425–34.

Liu, X., Yang, L., and Ren, J. 2008. Study on chip forming and breaking process based on virtual reality. *Proceedings of ICALIP* 1738–42.

Montgomery, D.C. 2005. *Design and Analysis of Experiments*. New York: John Wiley & Sons.

Wan, C.B.L., Liu, J.H., Ning, R.X., and Zhang, Y. 2006. Research and realization on CAD model transformation interface for virtual assembly. *Journal of System Simulation* 18: 391–94.

Weyrich, M., and Drews, P. 1999. An interactive environment for virtual manufacturing: the virtual workbench. *Computers in Industry* 38: 5–15.

Whyte, J., Bouchlaghem, N., Thorpe, A., and McCaffer, R. 2000. From CAD to virtual reality: Modelling approaches, data exchange and interactive 3D building design tools. *Automation in Construction* 10: 43–55.

Chapter 5

A Semi-Integration System of CAD and Inspection Planning of Standard Manufactured Features

Emad S. Abouel Nasr,[1,2] Abdulrahman Al-Ahmari,[1] and Osama Abdulhameed[1]

[1]*Industrial Engineering Department, College of Engineering, King Saud University, Riyadh, Saudi Arabia*

[2]*Mechanical Engineering Department, Helwan University, Cairo, Egypt*

Contents

Abstract

The inspection process is one of the most important steps in manufacturing industries that ensures high-quality and nondefective products to the customer. The manual inspection task cannot provide the desired quality due to the involvement of the human element in the process. Thus, automated inspection provides a reliable solution to a large number of problems inherent in the manual inspection task and to compete in the rapidly growing market. The automation of the inspection process requires the development of Computer-Aided Inspection Planning (CAIP), which is essential for modern industries. CAIP is an integrated system of Computer-Aided Design (CAD) and Computer-Aided Inspection (CAI). The objective of the proposed integrated system is to provide an interface that allows the user to identify the high-level features, such as slot, pocket, and hole, and lower-level entities, such as edge, point, and so forth for automatic inspection. The proposed framework of the integrated system is composed of three main modules: the Automatic Features Extraction Module (AFEM), Computer-Aided Inspection Planning Module (CAIPM), and Coordinate Measuring Machine Module (CMMM). The purposes of this chapter are (1) to address the issue of integration of CAD and inspection planning in which the prismatic features of design with inspection are linked efficiently, (2) to generate inspection plans automatically, and (3) to execute inspection plans through the incorporation of CAIP into CMM using DMIS code programming. Finally, a case study is presented to demonstrate and validate the integrated system.

5.1 Introduction

The inspection process is one of the most important steps in manufacturing industries that ensures high-quality and nondefective products to the customer. The inspection process is a very important task because it verifies whether the manufactured part lies within the tolerance of the designed model. The manual inspection task cannot provide the desired quality due to the involvement of the human element in the process. There are many factors such as tiredness, workspace, and lack of concentration of the operator that degrade the performance of the process with time. In some cases, manual inspection is not possible because the part to be inspected is either very small or the production rate of the part is very high (Yaqoub 2010).

An important prerequisite for the development of a CAD/CAM integrated system is the ability to increasingly automate the manufacturing activities that currently depend greatly on the human interface. With the advance of computer technology, the development of computer tools for integrating CAD and CAIP will help solve these disadvantages of manual inspection (Fan and Leu 1998).

CAIP helps to automate all the steps or activities of the inspection plan in order to increase the inspection efficiency and effectiveness (Moroni and Polini 1998). The inspection process must be able to provide feedback regarding the inspection results to the design and manufacturing processes. The main information required by an inspection process can be categorized in two areas: the design process and inspection process planning of the part. The design process involves extraction of the CAD model entities such as vertices, edges, loops, and faces that are used to recognize the features. Feature recognition involves the identification and grouping of feature entities from a geometric model (Abouel Nasr and Ali Kamrani 2007). The design processes also provide the technical specification of the part, which is Geometrical Dimensioning and Tolerancing (GD&T). The information related to inspection process planning describes the information required for preparing the final inspection plan.

Dependable and efficient product inspection plays a key role in achieving the quality requirements and in absorbing the regular variations in product design in a competitive worldwide marketplace. A good candidate for playing this role is the Coordinate Measuring Machine (CMM), which has many advantages over the traditional inspection methods (Zhang 2002). Nevertheless, CMM inspection planning is a difficult activity, which normally is the responsibility of professionals (Limaiem and El-Maraghy 2000). There are many tasks in CMM inspection planning, for example, determination of part setup, probe orientation, probe accessibility analysis, collision-free probe path planning, arrangement of measurement sequences, and finally, the determination of the number and coordinates of touching points (Venkata and Rao 2009; Menq et al. 1992; Lim and Menq 1994). Also, CMM can facilitate hybrid inspection planning for CMM-based measurements by combining a laser line scanner and a touch trigger probe. The inspection features being measured are identified and constructed based on the part's CAD model. After that, a knowledge-based sensor selection approach is applied to choose the most suitable sensor for each specified inspection feature. Then, two inspection planning modules—laser scanning module and tactile probing module—are developed to plan the inspection process automatically for all the specified inspection features. Finally, the measured data are collected and evaluated for tolerance verification (Zhao et al. 2012).

The purpose of this chapter is to make a link between the design and inspection (planning and execution) to get an efficient integration between CAD and CAIP that depends on an information model of a product model and a process model, so that the relationship between them is direct. This enables an efficient link between design activity, planning activity, and inspection execution (Kamrani and Abouel Nasr 2008).

To solve the integration problem for CAD, CAIP, and NC, dimensional measurement equipment, the Coordinate Measurement Machine (CMM), will be used. CMM offers the same return to inspection that the NC machine offers to manufacturing. A well-integrated system improves flexibility, reduces the product life cycle, reduces setup time, minimizes operator error, improves accuracy, improves quality, and reduces cost (Marefat et al. 1993; Ahmad et al. 2001; Roy 1995; Fu et al. 2003; Hsin-Chi and Wen 2002; Huikange et al. 2002; Liu and Strong 2002).

5.2 Review of Prior Related Research

The integration of design and manufacturing is the most important key to an efficient concurrent engineering (CE) environment. Through this integration, more economical and higher-quality complex manufacturing products can be achieved (Lionel et al. 2003). The concept of feature extraction recently has become a topic of interest in the field of CAD because it has great impact on directly connecting CAD with CAM (Aslan et al. 1999; Chandra and Ghosh 1999; Devireddy and Ghosh 1999). The objective of the extraction feature is to extract the manufacturing features at a high level of abstraction from low-level entities such as vertices and lines.

Therefore, feature extraction is considered a link between CAD and CAM. CAD/CAM systems should contain data, algorithms, and a defined structure, which automatically generate the design, functions, service life, manufacturing methods, and all the data needed for processing customer orders. CAD/CAM systems are the key factor in determining the success of various product development strategies and industrial competitiveness (Peihua 1994). The effective integration of CAD/CAM occurs by combining the feature-based modeling and feature extraction and recognition techniques in a single framework (Lee and Kim 2003). Feature extraction or Feature-Based Design (FBD) is a product modeling approach that includes both geometry and nongeometry information in a product definition such as (Chang 1990): geometry, tolerances, form features, material characteristics, topology, surface finish, process specification, and other data related to the attributes of the product. There is a need to combine feature-based modeling and feature recognition to achieve integration in the design and the process planning (Tseng 1999). A feature recognition system based on an intelligent feature recognition methodology (IFRM) is developed that has the ability to communicate with various CAD/CAM systems (Abouel Nasr and Kamrani 2006), and a feature recognition processor is developed to recognize not only manufacturing features but also replicate, compound, and translate features (Li, Ong, and Nee 2003).

Extracting the manufacturing information from the present CAD system and automatically generating all the information needed for downstream activities such as process planning and inspection is difficult to achieve. The current state emphasis on product quality, availability, and cost has turned into a progressively important role played by inspection. Therefore, integrating CAD and CAIP systems of manufacturing products has been an important subject for research in the recent literature (Sreeram et al. 2004; Cayiroglu 2009). A hybrid knowledge-based approach integrates CAD and CAIP into computer-aided design and inspection planning (CADIP), the product-based reasoning strategies required to implement computer-aided inspection process planning (CAIPP; Wong et al. 2006).

An object-oriented planner for inspection of prismatic parts is developed (Beg and Shunmugam 2002). Two stages, global inspection planning and local inspection planning stages, are made to prepare the best inspection plan (Lee et al. 2004). In the global inspection planning stage, the system generates an optimum inspection sequence of the features by analyzing the feature information. Inspection planning is done by including a selection of the most stable part orientations, arriving at the number and distribution of inspection points, feature accessibility analysis, sequencing of probe orientations, removal of duplicate faces and, finally, sequencing of faces (Beg and Shunmugam 2002; Lee et al. 2004).

There are many tools in the current technology used for inspection activities; the application of CMM to dimensional inspection is extensively used in the industry. However, the flexibility and accuracy of the CMM are not fully achieved as CMM is still operated manually by human workers (Yau and Meno 1993). From the inspection plan is generated a high-level inspection plan (HLIP) and a low-level inspection (LLIP) (Steven 1999). Also, a method of CAD-directed inspection path planning for coordinate measuring machines which are applicable to any object whose boundary is composed of planar, cylindrical, and conical faces (Kuang and Ming, 1998).

From the literature, there have been many trials that have addressed the integration of numerical controlled measuring machines with CAD systems by the use of computer-aided inspection process planning. The outcome of the previous research leads to the generation of automated inspection process plans (Duffuaa and AI-Najjar 1997; Yong et al. 2001), to the development of algorithms to determine feasible probe orientations and accessibility of CMM (Beg and Shunmugam 2002), to the determination of optimum measuring points using CMM (Cho and Kim 1995), to the development of a link between CAD and CMM through automated inspection planning (Legge 1996; Moroni and Polini 1998), and, finally, to the development of a methodology for automatically defining the accessibility domain of measurement points and forming a set of clusters (Vafaeesefat and EI-Maraghy 2000).

The ideal integrated system of CAD/CAIP/CMM should be able to recognize the input information from designers, manufacturers, and inspectors. Then this system generates inspection characteristics with the aid of geometric information from the CAD system in order to automate the inspection planning based on inspection knowledge and rules developed in the system. Geometrical dimensional and tolerance inspection planning should be capable of determining plans and information for measuring the dimensions and tolerances of the manufacturing products. The automated planning system should have the ability to (Zhao et al. 2002)

- Find the manufacturing features of the designed products (high level)
- Determine the topology between the different features
- Determine the low-level entities such as edges to be measured
- Determine the possible probe locations and approach directions
- Minimize probing operations while achieving successful measurement of all entities
- Determine the boundary of all faces which will be datum
- Determine the boundary of all faces, which will be geometrical tolerance tests such as flatness, perpendicularity, and so forth.

Moreover, the system should, in an optimal way, be able to automate the process of determining the selection of critical functional features based on the design intent involved in the CAD database. This is achieved by developing an inspection system that removes this responsibility from the user or programmer (Ziemian and Medeiros 1997). This ability also has a significant effect on various inspection planning activities such as reducing measurement points, sequences and paths, traveling distances, and positions of measurement points.

5.3 The Proposed Methodology

The proposed integrated system is primarily designed for standardizing automatic inspection of prismatic parts. The function of the proposed system is to facilitate and provide an intelligent interface, which allows the user to recognize the high-level features, such as slot, pocket, and hole, and lower-level entities, such as edge and vertices, for inspection functions.

The standardized proposed methodology has a great impact on the manufacturing market all over the world. The proposed system has the ability to increasingly automate the manufacturing activities that currently depend greatly on the human interface. In the long run, the proposed methodology will reduce the product life cycle, improve product quality, and reduce manufacturing costs. In today's competitive manufacturing arena, one of the most important problems facing product distributors is controlling the quality and cost of products from suppliers. To come up with a solution, the market demands new tools that not only validate the product quality and predict vendor's cost, but also assist vendors to improve quality and to lower costs.

Automatic feature recognition from CAD solid systems greatly impacts the level of integration between CAD and CAM. CAD files contain detailed geometric information of a part that is not suitable for using in the downstream applications such as process planning and inspection. Different CAD or geometric modeling packages store the information related to the design in their own databases. The structures of these databases are different from one another. As a result, no common or standard structure has been developed so far that can be used by all CAD packages. For that reason, this chapter will propose an intelligent standardized feature recognition methodology to develop a feature recognition system that has the ability to communicate with various CAD/CAM systems. The proposed methodology is developed for 3D prismatic parts that are created by using a solid modeling package. The system takes as input a neutral file in the Initial Graphics Exchange Specification (IGES) format and then translates the information in that file to manufacturing information, which provides the proposed methodology the ability to communicate with various CAD/CAM systems. However, STEP can be used as another standard format. The boundary (B-rep) geometrical information of the part design is analyzed by a feature recognition program that is created specifically to extract the features from the geometrical information based on object-oriented and geometric reasoning approaches. The feature recognition algorithms are able to recognize the following features: slots (through, blind, and round corners), pockets (through, blind, and round corners), holes (blind and through), and steps (through, blind, and round corners). Also, the proposed methodology is developed to handle the interaction between the features. These features, called manufacturing information, are mapped to the process planning function as an application for CAM. The system is written in the C++ language on a PC-based system (Abouel Nasr and Kamrani 2006).

On the other hand, the Automatic Inspection Planning (AIP) module consists of several elements. The inspection attributes are to be developed while extracting the geometric information from the CAD model by the feature extraction module to develop the inspection plan based on the inspection knowledge and rules that are stored in the system. The inspection planning module will generate the inspection attributes, inspection points, and probing vectors for the extracted manufacturing features. The inspection point generated on the CAD model database can be considered as the target point, which the CMM is supposed to touch. On the other hand, the probing vector is the vector that is used for directing the probe tip toward the surface of the designed part. The proposed automatic inspection module framework is presented for generating automated inspection process plans based on the feature extraction module. The proposed framework is generative inspection planning for automated inspection, which is similar to the generative process planning for CAPP in manufacturing. In the integrated framework, the inspection plan is generated simultaneously with the actual design process of a product.

The framework demonstrated in Figure 5.1 shows explicitly the concept and methodology of the integrated system. The system is composed of three main modules:

- Automatic Features Extraction Module (AFEM).
- Automatic Inspection Planning Module (AIPM).
- Coordinate Measuring Machine Module (CMMM).

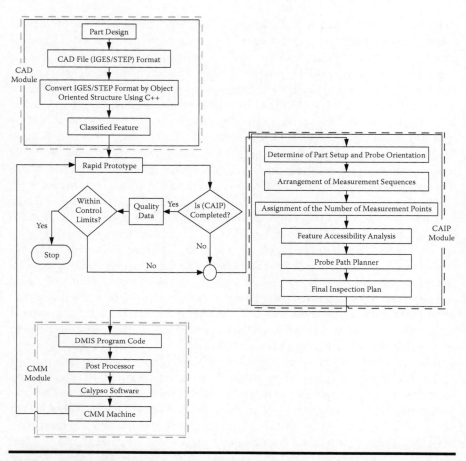

Figure 5.1   The proposed integrated system.

The integration system methodology is divided into three parts: the CAD module, the CAIP module, and the CMM module, which is described in the following discussion.

5.3.1 Automatic Feature Extraction Module (AFEM)

5.3.1.1 Feature Extraction from IGES File Format

The part design is introduced through CAD software and is represented as a solid model by using the Constructive Solid Geometry (CSG) technique as a design tool. The solid model of the part design consists of small and different solid primitives that combine to form the required part design. The CAD software generates and provides the geometrical information of the part design in the form of an ASCII file (IGES), the standard format; this provides the proposed methodology the ability to communicate with the various CAD/CAM systems. The boundary (B-rep) geometrical information of the part design is analyzed by a feature recognition program that is created specifically to extract the features from the geometrical information based on the geometric reasoning object-oriented approaches. The feature recognition program is able to recognize these features: slots (through, blind, and round corners), pockets (through, blind, and round corners), inclined surfaces, holes (blind and through), and steps (through, blind, and round corners). These features are called manufacturing information, which is mapped to process planning as an application for CAM. Finally, a rapid prototype will be made to compare features between the CAD model and the prototype model to verify the quality of the rapid prototype.

Feature extraction of prismatic parts created by using a Mechanical Desktop 6 power pack® CAD system that supports IGES file format translators at (B-REP Solid (186) with Analytical Surfaces). The feature recognition program is developed using Windows-based Microsoft Visual C++ 6 in a PC environment. The extracted entities are vertices, edges, loops, and faces. Feature recognition involves the identification and grouping of feature entities from a geometric model (Abouel Nasr and Kamrani 2007). Geometrical Dimensioning and Tolerancing (GD&T) information is also extracted from the CAD model, which is exported in an IGES file format. The IGES file format exports the GD&T as a general note by displaying all datum, all geometrical tolerance test types and their values, and the boundary of faces for every test and its datum. In the IGES file format, the geometrical tolerance in the MD6 design software, IGES translator, does not support tolerances but it is treated as a note in the exported IGES file.

Generally, faces are the basic entities that constitute the features, which are further defined by edges that are represented in terms of vertices, which are defined in terms of coordinates in the CAD file. Therefore, the hierarchy of the designed object represents the multilevels of different classes. All classes, except for the superclass representing an object as a whole, are objects of classes that are higher up in the data structure. For example, each edge object is represented in terms of vertex objects (Abouel Nasr and Kamrani 2006).

5.3.1.2 Feature Extraction from STEP File Format

Feature extraction of prismatic parts is created by using CATIA V5 R21 in a STandard for the Exchange of Product data (STEP), which is a product model data exchange standard (identified as ISO 10303). The STEP file format will be exported at AP203, which deals with configuration-controlled 3D designs of mechanical parts and assemblies (ISO10303-203:1994), which is one of the most widely used application protocols of STEP (Moroni and Polini 1998). The feature recognition program is developed using Windows-based Microsoft Visual C++ 6 in a PC environment. AP-203 Edition 2 is a recently released new version of the AP-203 standard for exchanging 3D geometry between CAD systems, and one of the extensions that it includes is GD&T data (Moroni and Polini 1998; Venkata and Rao, 2009).

For each feature shown in Figure 5.2, there is a specific production rule that defines how the feature will be extracted. For example, the following is the algorithm used for extraction of the slot blind feature, which is shown in Figure 5.3.

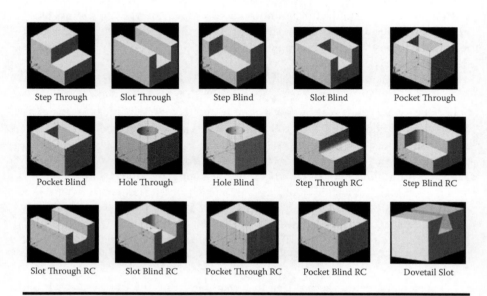

Step Through	Slot Through	Step Blind	Slot Blind	Pocket Through
Pocket Blind	Hole Through	Hole Blind	Step Through RC	Step Blind RC
Slot Through RC	Slot Blind RC	Pocket Through RC	Pocket Blind RC	Dovetail Slot

Figure 5.2 Manufacturing features. (From Abouel Nasr, E., and Kamrani, A.K. 2007. *Computer-Based Design and Manufacturing: An Information-Based Approach.* **New York: Springer.)**

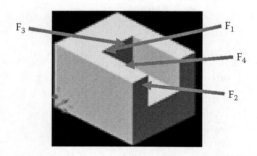

Figure 5.3 SLOT BLIND. (From Abouel Nasr, E., and Kamrani, A.K. 2007. *Computer-Based Design and Manufacturing: An Information-Based Approach.* **New York: Springer.)**

Rule: Feature SLOT BLIND

1. For every four faces of surface type plane (named as face 1 to face 4).
2. If face 1, face 2 concave edge count of outer loop equals 3 each and face 3, face 4 concave edge count of outer loop equals 2 each, and face 1 connected to face 2 by one edge.
   a. If face 1 perpendicular to face 2, face 3, and face 4 also face 3 and face 4 parallel to each other.
      i. SLOT BLIND found
      ii. Create a new SlotB object and add to feature list.
3. End For

The following steps are the proposed methodology for features extraction and classification:

Step 1: Extract the geometry and topology entities for the designed object model from IGES file or STEP file format.
Step 2: Extract topology entities in each basic surface and identify their type.
Step 3: Test the feature's existence in the basic surface based on loops.
Step 4: Identify the feature type.
Step 5: Identify the detailed features, and extract the related feature geometry parameters.
Step 6: Extract all GD&T test faces that depend on the functionality of the part.
Step 7: Identify the detailed machining information for each feature and the designed part.

5.3.2 Development of Automatic Inspection Module (AIM)

The function of this module is to help in accomplishing flexible dimensional inspection of prismatic parts by using CMM, as shown in Figure 5.4. The proposed integrating system would be developed to generate the inspection plans and comprehensive instructions for inspecting the final manufactured product. The system modules use a common database that connects and integrates their data. In inspection planning, the approach to measuring designed part attributes that contain both high-level features such as slots, pockets, holes, and so forth, and low-level features such as edges and vertices, will be realized by taking into consideration the relationships between the extracted features. The methodology used in developing the system is

▪ Using rule, structure, and pointer-based representations of knowledge by using an object-oriented approach.

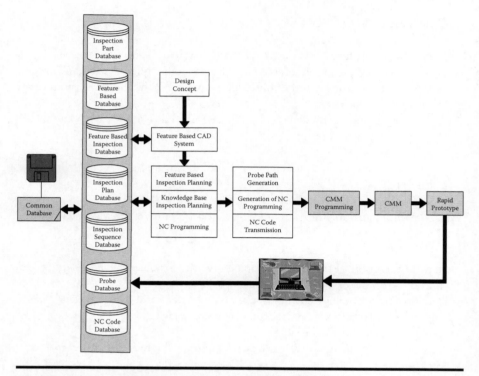

Figure 5.4 The proposed automatic inspection module.

- Developing data models and interfaces to establish the integrating system.
- Developing a knowledge-based geometric reasoning approach for automating inspection planning.

5.3.2.1 Database Module

To automate and integrate CAD and AIM, a structured database should be created, as shown in Figure 5.5. Eight elements are generated by the CAD system: object, feature, face, edge, starting vertex, terminating vertex, coordinate, and inspection specifications (Ketan et al. 2002). The object properties start from a designed part that is linked to its extracted feature group. Each feature is linked to its face group, then linked to its group of edges in the edge group, which is linked to its group of vertices in the starting and terminating vertex group. At the end, each feature is linked to its inspection specification group. The database can be extended to include three groups: criteria group, setting group, and dummy active group, which are used as the main data to generate the inspection planning instructions.

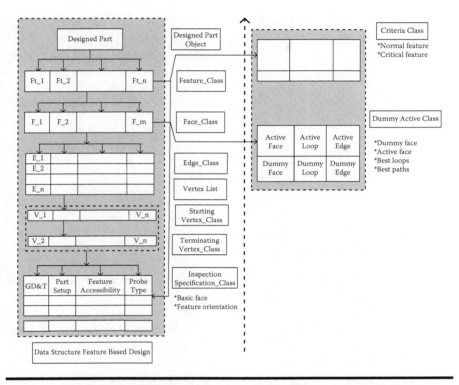

Figure 5.5   The proposed data structure of the automatic inspection module.

5.3.2.2 Developing the Integration between CAD and CAI

The database of any CAD system for a designed part contains low-level entities such as points, lines, and so on. This geometric information is not suitable for downstream applications. The feature extraction approach is considered an efficient way to recognize design part geometry, tool geometry, tool motion path, probe motion path, and features that have all the necessary information for process planning. By developing a good database for the designed part, a computer integrated manufacturing environment can be achieved.

Manufacturing features can be categorized into simple features and complex features. Complex features can be defined by the interaction of two or more simple features. The basic element of any feature is the face in which the feature is created. This face can be identified by basic face (BF). After establishing the detailed database of the extracted features from the CAD system, the inspection process tasks and their sequences are identified. By using a CNC machine and Rapid Prototype machine, a detailed inspection plan can be developed by determining the inspection operations and sequencing them. The basic problem that faces recent approaches to integration of CAD and CAIP is the user interface, in which an automatic design data extraction is achieved without the need to extract technological and manufacturing information manually. The feature extraction module will play an important link in eliminating the need to apply an interface between design and manufacturing.

This chapter will use an object-oriented approach to solve some of the problems listed before by developing an intelligent feature extraction system and inspection planning system. This module will contain a computer database for a designed part that has all the information needed for direct generation of CNC code for the inspection task. The proposed system will integrate object-oriented inspection planning with the CAD system for automatic CNC and Rapid Prototype Machine measuring program generation of prismatic parts.

5.3.2.3 Generating the Inspection Plan for the Manufacturing Components

Inspection process activities and their sequences are determined based on the feature extraction module from CAD data. Concerning the CNC-controlled measurement machine and Rapid Prototype machine as in this chapter, it is necessary to detail the inspection plan by determining the inspection operations and sequencing them. The main obstacle facing current techniques of integration between CAD and CAIP is an automatic and direct design data interpretation without the need to extract technological and manufacturing information. The proposed feature extraction methodology had opened new frontiers in eliminating the need to apply an interface between the two activities and to provide an efficient processing of data knowledge to support downstream activities from the design stage.

This research introduces a knowledge-based methodology to resolve a number of previous problems by integrating computer-aided design/inspection planning system. This system contains a computer-internal model of a product that contains sufficient information to guide a direct generation of CNC code for inspection activity. The developed small prototype system integrates knowledge-based inspection planning with a feature-based CAD system for automatic CNC measuring program generation of prismatic parts. Two concepts are proposed to prepare the inspection knowledge of the feature families being built: first, dummy and active face feature faces and, second, primary and secondary feature faces.

A successful inspection planning approach for planer and cylindrical surfaces of prismatic manufacturing features by using a coordinate measuring machine (CMM) will be generated. The inspection planning activity is based on critical inspection specifications rather than tracing all dimensions on the design part as an essential concept. Therefore, this approach provides the best possible measuring points and path generation of the CMM probe. On the other hand, by choosing the probing sequence of measuring points in optimal way, probe movement and inspection time can be minimized. Thus, a probe-path generation method proposed in this chapter will lead to minimize inspection time (Adil and Ketan 2005).

The inspection planning module is a knowledge based system, as shown in Figure 5.6. The preparation of inspection knowledge entails the following items:

- Working faces (BF) in which features are created
- High-level feature types created on a certain (BF)
- Feature orientation and probe location (S)
- Setting for each feature, which is determined by feature type and orientation
- Inspection parameters and measuring edges of a feature for each setting
- Edge limits and edge values
- Probing parameters (i.e., probe approach direction and probe inspection direction)

A major component of inspection planning module is the inspection database, which includes declarative database and procedural database. Declarative database is information about the part and its features, inspection characteristics, specifications, manufacturing processes, and so on. Both declarative and procedural databases can be considered as a system's problem-solving knowledge (Adil and Ketan 2005). In the generation of the inspection plan, there are some settings required before building the inspection plan table.

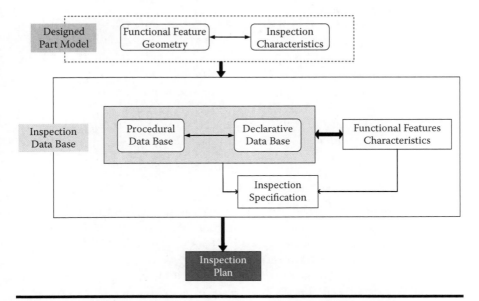

Figure 5.6   The proposed hierarchy of the inspection module.

5.3.2.4 The Best Orientation of the Prismatic Parts

To get the best orientation of the prismatic parts, the following methodology explains the steps to get the best orientation:

- A list of all faces that are contained in all features will be created.
  - In this step, a list of all faces that are contained in all features will be created to compare them with another list.
- A list of all faces that are contained in the part will be created.
  - In this step, a list of all faces of the whole part will be listed to compare between them and the list of all faces of all features.
- All intersecting faces between features and all faces of the part will be filtered.
  - In this step, all intersections between the list of faces of all features and the list of faces of the whole part will be filtered.
- The other faces will be taken, among which one of them will be the bottom face at fixing the part on the CMM. One of the resulting faces after filtering the intersection between the list of faces of all features and the list of faces of the whole part will be the best bottom face to fix the part in the CMM table.
- The filtering faces will be arranged depending on the less intersection of edges between them and the edges of the all faces of all features.
- The comparison of these faces with one another will be made. If one of these faces is in the same plane with one or more faces, the count of the intersection's edges of every face that are in the same plan will be add to the last summation for every face.
- These faces will be rearranged depending on the less intersection of edges where less intersection is better.
- The first one that is the least intersection summation will be the best bottom face in the first setup.
- If all the features are accessible by probe selection at this setup, the setup methodology will be stopped. If not, go to the next setup.
- The next face sequence that is not in the same plan from the list of less intersection's edges will be the bottom of the next setup.

5.3.2.5 Probe Accessibility Analysis

The determination of the probe accessibility analysis for every feature will be automatically generated. The probe accessibility analysis represents the accessibility direction of the probe to measure the feature. For example, a slot blind as shown in Figure 5.7 has only two probe directions.

Then, the normal vector for every face of the feature will be taken to find the accessibility of the probe to the slot blind feature.

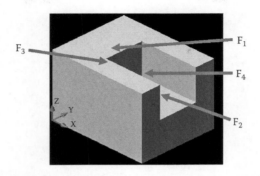

Figure 5.7   Step blind feature.

Table 5.1   Clustering Algorithm Analysis for Step Blind Feature

	Normal Vector	*(+X)*	*(−X)*	*(+Y)*	*(−Y)*	*(+Z)*	*(−Z)*
Face Id# 1	(−1,0,0)	0	1	0	0	0	0
Face Id# 2	(1,0,0)	1	0	0	0	0	0
Face Id# 3	(0,−1,0)	0	0	0	1	0	0
Face Id# 4	(0,0,1)	0	0	0	0	1	0
Sum		1	1	0	1	1	0
Accessibility		0	0	1	0	0	1

The probe approach direction for the slot blind feature will be from the +y and −z directions. Finally, a clustering algorithm will be applied to find the best selection of the type of the probe at the part that has more than one feature as shown in Table 5.1.

Table 5.2 shows the inspection plan, which has all the required information as follows:

- Face ID, which is the ID number of the face in the features extraction and recognition output.
- The ID of the inspection operation, which is the type of the GD&T, such as flatness, perpendicularity, and so on.
- Tolerance value, which allows the designers to set tolerance limits for all of the various critical characteristics of a part by examining its function and its relationship to mating parts.

Table 5.2   Inspection Plan

Face ID	ID of Inspection Operation	Tolerance Value	Tool Used	Datum Faces ID	Orientation of Part	No. of Touch Point	Coordinates of Touch Point	Geometric Inspection Boundary
1	GD&T	Tolerance value	Probe Type	Face ID	Setup No.	n	$(X_i, Y_i, Z_i)-$ $(X_i + 1, Y_i + 1, Z_i + 1)$ $(X_i + 2, Y_i + 2, Z_i + 2)$ . . $(X_n, Y_n, Z_n) -$	$(X_1, Y_1, Z_1) -$ $(X_2, Y_2, Z_2) -$ $(X_3, Y_3, Z_3) -$ $(X_4, Y_4, Z_4) -$

- Tools used for inspection such as horizontal probe, vertical probe, and star probe, the selection of which depends on the location, depth, dimensions, and orientation of the features and also the radius, and length of the probe to avoid collisions between the part and the probe.
- Datum faces that are the reference points, lines, planes, and axis, which are assumed to be exact.
- Orientation of the part, which is the best orientation of the prismatic part and the setups number of the part to cover all the required features.
- Number of touch points (n), which will be fixed at five points.
- Touch point, which is the coordinate of the probing point.
- Boundary of the testing face.

5.3.3 Coordinate Measuring Machine Module (CMMM)

By using a coordinate measuring machine (CMM), a successful inspection planning methodology for simple manufacturing features will be developed. The inspection planning methodology is based on a critical inspection specification rather than checking all dimensions of the designed part. Moreover, the module will have the ability to select the optimal measuring points and path generation of the CMM probe. Therefore, by selecting the probing sequence of measuring points optimally, probe movement and inspection time can be reduced. As a consequence, the probe path generation method proposed in this chapter will lead to the minimization of inspection time and, subsequently, the product cost.

The inspection planning that is generated using the CAIP module is imported to CMM by generation DMIS, which is a programming language used for programming mechanical, optical, laser, and video measuring systems. DMIS files must be translated to ensure that they can also be used by measuring machines that are not DMIS-compatible. The translator tailored to relevant measuring software is referred to as the post processor, because it becomes active after the production of the DMIS file. The typical lines of the DMIS file format for a plane are shown in Figure 5.8.

5.4 A Case Study

The proposed methodology is used for the component illustrated by Figures 5.9a, 5.9b and 5.9c. Mechanical Desktop 6 Power Pack is the CAD system used that supports B-rep and IGES translator. However, another similar CAD system that supports the IGES translator can be used. The proposed methodology is developed by using Microsoft Visual C++ 6 in a Windows-based PC environment. The designed object consists of seven different features and prismatic raw material. By applying the proposed methodology, parts of the analyzed results are shown in Tables 5.3 and 5.4.

```
DMISMN/'OSAMAMEASU'
FILNAM/'OSAMAMEASU.DMO'
UNITS/MM, ANGDEC, TEMPC
DMISMN/'DMIS Program'
......
MEAS/PLANE, F(FRONTPLANE), 5
GOTO/–8.6927, –1.4262, –5.8781
PTMEAS/CART, –8.6906, 0.0760, –5.8781, –0.0000, –1.0000, 0.0000
GOTO/–10.3712, –1.4162, –30.1770
PTMEAS/CART, –10.3690, 0.0859, –30.1771, –0.0000, –1.0000, 0.0000
PTMEAS/CART, 70.5000000000, 40.0000000000, –18.0000000000, $
–1.0000000000, 0.0000000000, 0.0000000000
......
$$ Changing label 'Right Slot Flatness' to: 'RIGHTSLOTF'.
T(RIGHTSLOTF)=TOL/FLAT, 0.0100000
OUTPUT/FA(RIGHTSLOTP), TA(RIGHTSLOTF)
$$ Changing label 'Perpendicularity1' to: 'PERPENDICU'.
T(PERPENDICU)=TOL/PERP, 0.0200000, RFS, F(SLOTLEFTPL)
OUTPUT/FA(TOPMIDLEPL), TA(PERPENDICU)
ENDFIL
```

Figure 5.8   DMIS file.

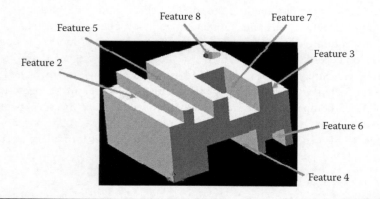

Figure 5.9a   An illustrative example (solid).

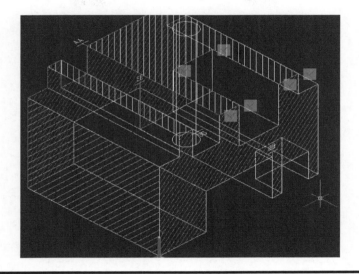

Figure 5.9b   An illustrative example (datum faces).

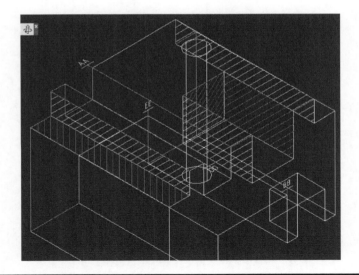

Figure 5.9c   An illustrative example (geometrical tolerance faces).

Table 5.3 Manufacturing Features

Feature ID	Faces ID	Common Edges ID	Location	Feature Name	Dimension			
					L	W	H	R
[1]			[24] = (0,0,0)	Raw Material	120	100	80	
[2]	[2][3]	[5]	[6] = (107.5,0,60)	Step_Through	100	20	12.5	
[3]	[6][7]	[36]	[22] = (12.5,0,65)	Step_Through	100	15	12.5	
[4]	[15][16][17]	[55][58]	[26] = (35,0,60)	Slot_Through	100	50	30	
[5]	[23][24][25]	[71][74]	[19] = (25,0,60)	Slot_Through	100	20	20	
[6]	[10][11][12][13]	[47][48][50][45][51]	[38] = (92.5,20.0)	Slot_Blind	20	20	20	
[7]	[19][20][21][22]	[63][64][66][67][69]	[15] = (60,0,50)	Slot_Blind	30	30	50	
[8]	[1][9]	[1][2][3][4][43][44]	[29] = (95,75,0)	Hole_Through				7.514

Table 5.4 Machining Information

Operation Sequence	Feature ID	Feature Type	Operation Type	Machine	Cutting Tool	Tool Approach	Removed Volume
1	[2]	Step_Through	Shoulder_Milling	Milling	Side Milling Cutter	[0,1,0] or [0,−1,0]	25000.00
2	[3]	Step_Through	Shoulder_Milling	Milling	Side Milling Cutter	[0,1,0] or [0,−1,0]	18750.00
3	[5]	Slot_Through	Slotting_Milling	Milling	End Milling Cutter	[0,1,0] or [0,−1,0]	40000.00
4	[7]	Slot_Blind	Slotting_Milling	Milling	End Milling Cutter	[0,0,−1] or [0,1,0]	45000.00
5	[4]	Slot_Through	Slotting_Milling	Milling	End Milling Cutter	[0,1,0] or [0,−1,0]	15000.00
6	[6]	Slot_Blind	Slot_Blind	Slotting_Milling	Milling	[0,0,1] or [0,1,0]	8000.00
7	[8]	Hole_Through	Drilling	Drilling	Twist Drill	[0,0,−1] or [0,0,−1]	14137.17

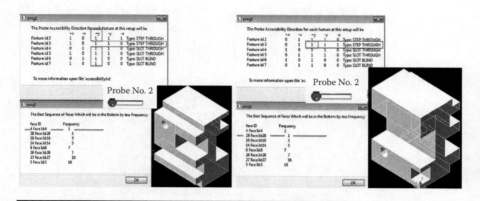

Figure 5.10 Part setups and orientation.

All the extracted manufacturing features in terms of the feature identification number (ID), feature name, feature dimensions, and features' location relative to the original coordinates of the deigned part are shown in Table 5.3. The machining information of the designed part that includes the machining sequence, the operation type, the machine, the cutting tool, the tool/machining approach, and the removed volume for each extracted feature are summarized in Table 5.4.

The best face setup that allows the probe to access all or most features is shown in Figure 5.10. Finally, the generated inspection plan of the case study is shown in Table 5.5. Moreover, the output graph of the flatness is shown in Figure 5.11, which displays the deviation of the points.

5.5 Conclusion

The proposed standardized integrated system has been developed primarily for standardizing the automatic inspection of prismatic parts. The objective of the proposed system is to help provide an intelligent interface that permits the user to recognize high-level features, such as slot, pocket, and hole, and lower-level entities such as edge, vertices, and so forth, for inspection functions. The standardized proposed methodology has an enormous impact on the manufacturing market all over the world. The proposed system has the ability to increasingly automate the manufacturing activities that currently depend heavily on the human interface. The proposed methodology will reduce the product life cycle, improve product quality, and reduce the manufacturing cost.

Acknowledgments

This works is funded by National Plan for Science & Technology (NPST), Saudi Arabia.

Table 5.5 Inspection Planning for the First and Second Setup of the Part

Face ID	ID of Inspection Operation	Tolerance Value	Tool Used	Datum Faces ID	Setup of the Part	No. of Touch Point	Coordinates of Touch Point	Geometric Boundary of Inspection Face
1	▱	0.066	Horizontal Probe	AA, CC, EE	First Setup	5	(90,3,53)	(90,0,50)
							(90,3,77)	(90,0,80)
							(90,47,77)	(90,50,80)
							(90,47,53)	(90,50,50)
							(90,6,56)	
2	⊥	0.044	Horizontal Probe	AA, BB, DD	First Setup	5	(90,3,53)	(90,0,50)
							(90,3,77)	(90,0,80)
							(90,47,77)	(90,50,80)
							(90,47,53)	(90,50,50)
							(90,6,56)	

(Continued)

Table 5.5 (Continued) Inspection Planning for the First and Second Setup of the Part

Face ID	ID of Inspection Operation	Tolerance Value	Tool Used	Datum Faces ID	Setup of the Part	No. of Touch Point	Coordinates of Touch Point	Geometric Boundary of Inspection Face
3	▱	0.082	Horizontal Probe	AA, CC, EE	First Setup	5	(60,3,53)	(60,0,50)
							(60,3,77)	(60,0,80)
							(60,47,77)	(60,50,80)
							(60,47,53)	(60,50,50)
							(60,6,56)	
4	⊥	0.077	Horizontal Probe	AA, BB, DD	First Setup	5	(60,3,53)	(60,0,50)
							(60,3,77)	(60,0,80)
							(60,47,77)	(60,50,80)
							(60,47,53)	(60,50,50)
							(60,6,56)	
5	▱	0.01	Horizontal Probe	AA, BB, CC	Second Setup	5	(15,3,68)	(15,100,65)
							(15,3,77)	(15,100,80)
							(15,97,77)	(15,0,80)
							(15,97,68)	(15,0,65)
							(15,6,71)	

6	╲	0.05	Horizontal Probe	AA, BB, CC	Second Setup	5	(15,3,68)	(15,100,65)
							(15,3,77)	(15,100,80)
							(15,97,77)	(15,0,80)
							(15,97,68)	(15,0,65)
							(15,6,71)	
7	▱	0.01	Horizontal Probe	AA, BB, CC	Second Setup	5	(3,3,65)	(0,0,65)
							(3,97,65)	(0,100,65)
							(12,97,65)	(15,100,65)
							(12,3,65)	(15,0,65)
							(6,6,65)	
8	⊥	0.05	Horizontal Probe	AA, CC, EE	Second Setup	5	(63,50,53)	(60,50,50)
							(63,50,77)	(60,50,80)
							(87,50,77)	(90,50,80)
							(87,50,53)	(90,50,50)
							(66,50,56)	

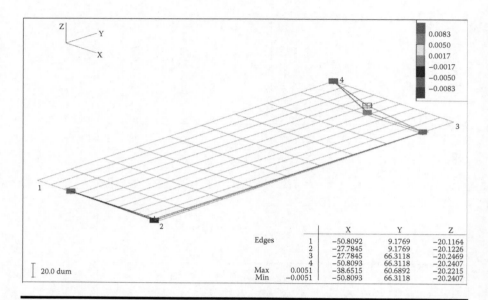

Edges		X	Y	Z
	1	−50.8092	9.1769	−20.1164
	2	−27.7845	9.1769	−20.1226
	3	−27.7845	66.3118	−20.2469
	4	−50.8093	66.3118	−20.2407
Max	0.0051	−38.6515	60.6892	−20.2215
Min	−0.0051	−50.8093	66.3118	−20.2407

Figure 5.11   Flatness graph of the plane of the slot feature.

Authors

Emad S. Abouel Nasr is an Associate Professor in Industrial Engineering Department, College of Engineering, King Saud University, and Assistant Professor in Mechanical Engineering, Department at Helwan University, Faculty of Engineering, Helwan, Cairo, Egypt. He received his PhD in Industrial Engineering from the University of Houston, Texas, USA, in 2005. His current research focuses on CAD, CAM, rapid prototyping, advanced manufacturing systems, and collaborative engineering.

Abdulrahman Al-Ahmari is a Professor of Industrial Engineering at King Saud University, Saudi Arabia, Dean of Advanced Manufacturing Institute. He received his PhD (Manufacturing Systems Engineering) in 1998 from the University of Sheffield, the United Kingdom. His research interests are in analysis and design of manufacturing systems; computer integrated manufacturing (CIM); optimization of manufacturing operations; applications of simulation optimization; FMS, DOE, and cellular manufacturing systems.

Osama Abdulhameed is a master's student of Industrial Engineering at King Saud University, Saudi Arabia. He received his BSc (Manufacturing Systems Engineering) in 2009 from King Khalid University. His research interests are in developing an integrated system for CAD and inspection planning.

References

Abouel Nasr, E., and A.K. Kamrani. 2006. A new methodology for extracting manufacturing features from CAD system. *International Journal of Computer and Industrial Engineering* 51: 389–415.

Abouel Nasr, E., and Kamrani, A.K. 2007. *Computer-Based Design and Manufacturing: An Information-Based Approach*. New York: Springer.

Adil, M.B., and Ketan, H.S. 2005. Integrating design and production planning with knowledge-based inspection planning system. *The Arabian Journal for Science and Engineering* 30(2B).

Ahmad, N., Haque, A.F.M., and Hasin, A.A. 2001. Current trend in computer aided process planning. *Proceedings of the 7th Annual Paper Meeting and 2nd International Conference of the Institution on Engineering* 10: 81–92.

Aslan, E., Seker, U., and Alpdemir, N. 1999. Data extraction from CAD model for rotational parts to be machined at turning centers. *Turkish Journal of Engineering and Environmental Science* 23(5): 339–47.

Beg, J., and Shunmugam, M.S. 2002. An object oriented planner for inspection of prismatic parts—OOPIPP. *International Journal of Advanced Manufacturing Technology* 19: 905–16.

Cayiroglu, I. 2009. A new method for machining feature extracting of objects using 2D technical drawings. *Computer-Aided Design* 41: 1008–19.

Chandra, R.D., and Ghosh, K. 1999. Feature-based modeling and neural networks-based CAPP for integrated manufacturing. *International Journal of Computer Integrated Manufacturing* 12(1): 61–74.

Chang, T.C. 1990. *Expert Process Planning for Manufacturing*. Boston: Addison-Wesley Publishing Company.

Cho, M.W., and Kim, K. 1995. New inspection planning strategy for sculptured surface using coordinate measuring machine. *International Journal of Production Research* 33(2): 422–44.

Devireddy, C. R., and Ghosh, K. 1999. Feature-based modeling and neural networks-based CAPP for integrated manufacturing. *International Journal of Computer Integrated Manufacturing* 12(1): 61–74.

Duffuaa, S.O., and AI-Najjar, H.J. 1997. A general inspection plan for critical multi-characteristic components. *International Journal of Production Research* 35(10): 2723–36.

Fan, K. C., and Leu, M.C. 1998. Intelligent planning of CAD-directed Inspected for coordinate measuring machine. *Computer Integrated Manufacturing System* 11(1–2): 43–51.

Fu, M.W., Ong, S.K., Lu, W.F., Lee, I.B.H., and Nee, A.Y.C. 2003. An approach to identify design and manufacturing features from a data exchanged part model. *Computer Aided Design* 35(1): 979–93.

Hsin-Chi, C., and Wen, F.L.U. 2002, Machining process planning of prismatic parts using case-based reasoning and past process knowledge. *Applied Artificial Intelligence* 16: 303–31.

Huikange, K., Nandakumar, M., and Shah, J. 2002. CAD/CAM integration using machining features. *International Journal of Computer Integrated Manufacturing* 15(4): 296–318.

Kamrani, A.K., Abouel Nasr, E., and Kanawade, M. 2008. Product design and development framework in collaborative engineering environment. *International Journal of Computer Applications in Technology* 32(2): 85–94.

Ketan, H.S., Al-Bassam, M.A., Adel, M.B., and Rawabdeh, I. 2002. Modeling integration of CAD and inspection planning of simple element features. *Integrated Manufacturing Systems* 13(7): 498–511.

Kuang-Chao, Fan and Ming, C. Luet. 1998. Intelligent Planning of CAD-directed inspection for coordinate measuring machines. *Computer Manufacturing System* II (1–2), 43–51.

Lee, K. 1999. *Principles of CAD/Cam/CAE Systems*. Boston: Addison Wesley.

Lee, J.Y and Kim.K. 1998. A feature-based approach to extracting machining features. *Computer-Aided Design*, 30(13), 1019–1035.

Lee, H., Cho, M.W., Gil-Sang Yoon, G.S., and Choi, J. 2004. A computer-aided inspection planning system for on-machine measurement-Part I: Global inspection planning. *KSME international Journal* 18(8): 1349–57.

Legge, D.I. 1996. Integrating design and inspection systems: A literature review. *International Journal of Production Research* 34(5): 1221–41.

Li, W.D., S.K. Ong, Nee, A.Y.C. 2002, Recognizing manufacturing features from a design by feature model. *Computer Aided Design* 34 P: 849–868.

Lionel, R., Otto, S., and Henri, P. 2003. Process planning as an integration of knowledge in the detailed design phase. *International Journal of Computer Integrated Manufacturing* 16(1): 25–37.

Lim, C.P., and Menq, C.H. 1994. CMM feature accessibility and path generation. *International Journal of Production Research* 32(3): 597–618.

Limaiem, A., and El-Maraghy, H.A. 2000. Integrated accessibility and measurement operations sequencing for CMMs. *Journal of Manufacturing Systems* 19(2): 83–93.

Liu, J.X., and Strong, R. 2002. Machining fixture verification for linear fixture systems. *International Journal of Production Research* 40(14): 3441–59.

Marefat, M., Malhorta, S., and Kashyap, R.L. 1993. Object oriented intelligent computer integrated design, process planning, and inspection. *IEEE Computer.* 54–65, Vol. 26.

Menq, C.H., Yau, H,-T., and Wong, C.L. 1992. An intelligent planning environment for automated dimensional inspection using CMMs. *Journal of Engineering for Industry, Transactions of the ASME* 114: 222–30.

Moroni, G., and Polini, W. 1998. Form feature selection in computer aided inspection planning. *International Journal of Flexible Automation and Integrated Manufacturing* 6(1,2): 29–51.

Peihua, G. 1994. Feature representation scheme for supporting integrating manufacturing. *Computers and Industrial Engineering* 26(1): 55–71.

Roy, U. 1995. Computational methodologies for evaluating form and positional tolerances in a computer integrated manufacturing system. *International Journal of Advanced Manufacturing Technology* 10(2): 110–117.

Sreeram, S., Acharya, B., and Sam, S. 2004. A factor analysis approach for robust inspection of circular features with lobing errors. *Transactions of the North American Manufacturing Research Institute of SME* 32: 135–42.

Steven, N.S. 1999. Dimensional Inspection Planning for Coordinate Measuring Machines. Ph.D. diss., University of Southern Carlifornia.

STEP application handbook ISO 10303: Version 3, 30 June 2006.

Tseng, Y.J. 1999. A modular modeling approach by integrating feature recognition and feature-based design. *Computers in Industry* 39: 113–25.

Wong, F.S.Y, Chuah, K.B., and Venuvinod, P.K. 2006. Automated inspection process planning: Algorithmic inspection feature recognition, and inspection case representation for CBR. *Robotics and Computer-Integrated Manufacturing*, 22: 56–68.

Vafaeesefat, A., and EI-Maraghy, H.A. 2000. Automated accessibility analysis and measurement clustering for CMMs. *International Journal of Production Research* 38(10): 2215–31.

Venkata Bhaskar Sathi, S., and Rao, P.V.M. 2009. STEP to DMIS: Automated generation of inspection plans from CAD data. *5th Annual IEEE Conference on Automation Science and Engineering*, Bangalore, India, August 22–25.

Yaqoub, Z.H. 2010. Feature-Based Integration of Design/Process/Inspection Planning for Rotational Parts. Ph.D. diss., University of Technology, Iraq.

Yau, H.T., and Meno, C.H. 1993. Concurrent process planning for finish milling and dimensional inspection of sculptured surfaces in die and mold manufacturing. *International Journal of Production Research* 31(11): 2709–25.

Yong, K., Wang, E., and Hyung, M.R. 2001. Geometry-based machining precedence reasoning for feature-based process planning. *International Journal of Production Research* 39(10): 2077–103.

Zhang, G.X., Liu, S.G., Ma, X.H., and Wang, Y.Q. 2002. Toward the intelligent CMM. *CIRP Annuals-Manufacturing Technology* 51(1): 437–42.

Zhao, H., Kruth, J., Gestel, N.V., Boeckmans, B., and Bleys, P. 2012. Automated dimensional inspection planning using the combination of laser scanner and tactile probe. *Measurement* 45: 1057–66.

Zhao, Y., Ridgway, K., and Al-Ahmari, A.M.A. 2002. Integration of CAD and a cutting tool selection system. *Computers and Industrial Engineering* 42: 17–34.

Ziemian, C.W., and Medeiros, D.J. 1997. Automated feature accessibility algorithm for inspection on a coordinate measuring machine. *International Journal of Production Research* 35(10): 2839–56.

Chapter 6

Tumor Geometrical Deformation Modeling

Maryam Azimi,[1] Ali K. Kamrani,[2] and Emad Samir Abdelghany[3]

[1]Lenovo Corporation, Morrisville, North Carolina

[2]Industrial Engineering Department, University of Houston, Houston, Texas

[3]Industrial Engineering Department, College of Engineering, King Saud University, Riyadh, Saudi Arabia

Contents

Abstract

Many types of cancer have been treated by radiation therapy in recent years. The treatment objective is to achieve tumor control by planning a significant total dose of radiation to the cancerous region to sterilize the tumor without damaging the surrounding healthy tissues. Most of the current radiation treatment planning systems use pre-treatment CT images to design targets and assume that the tumor geometry will remain the same throughout the treatment period, which takes 5 to 7 weeks. However, this assumption is flawed because the tumor geometry has been shown to change over time. As a result, the treatment may be suboptimal. There is a critical need to understand how tumor geometry changes over time during the radiation treatment. This chapter presents the results of an ongoing research in the development of a three-dimensional (3D) prediction model for tumor deformation during radiation treatment for cancer patients. The rationale for this project was to successfully implement a framework for treatment planning using predictive models and to increase the accuracy of the treatments.

6.1 Introduction

Cancer is a group of related diseases that begin in a cell, the body's basic unit of life. To better understand cancer, it is helpful to know what happens when normal cells become cancerous. The body is made up of many types of cells. Normally, cells *grow* and divide to produce more cells only when the body needs them. This orderly process helps keep the body healthy. Sometimes, however, cells keep dividing when new cells are not needed. These extra cells form a mass of tissue called a growth or *tumor.* Tumors can be benign or malignant. They can often be removed, and, in most cases, they do not return. Cells from benign tumors do not spread to other parts of the body. Most important, benign tumors are rarely a threat to life. Malignant tumors are called cancer. Cells in these tumors are abnormal and divide without control or order. They can invade and damage nearby tissues and organs. Treatment for cancer depends on the type of cancer; the size, location, and stage of the disease; the patient's general health; and other factors. People with cancer are often treated by a team of specialists, which may include a surgeon, a radiation oncologist, a medical oncologist, and others. Most cancers are treated with surgery, radiation therapy, chemotherapy, hormone therapy, or biological therapy. Doctors may decide to use one treatment method or a combination of these methods. Research studies evaluate promising new therapies and answer scientific questions. The goal of such trials is to find treatments that are more effective in controlling cancer with fewer side effects. Treatment for cancer can be either local or systemic. Local treatments affect cancer cells in the tumor and the area near it. Systemic treatments travel through the bloodstream, reaching cancer cells all over the body. Surgery and radiation therapy are types of local treatment. Therefore, positioning accuracy is critical for radiation therapy. Chemotherapy, hormone therapy, and biological therapy are examples of systemic treatment.

Radiation therapy uses high-energy X-rays or charged particles to kill cancer cells. For some types of cancer, radiation therapy may be used instead of surgery as the primary treatment. Radiation therapy also may be given before surgery (neoadjuvant therapy) to shrink a tumor so that it is easier to remove during surgery. In other cases, radiation therapy is given after surgery (adjuvant therapy) to destroy any cancer cells that may remain in the area. Radiation also may be used alone, or along with other types of treatment, to relieve pain or other problems if the tumor cannot be removed. Radiation therapy can be in either of two forms: external or internal. Some patients receive both. External radiation comes from a machine that aims the rays at a specific area of the body. Most often, this treatment is given on an outpatient basis in a hospital or clinic. There is no radioactivity left in the body after the treatment. With internal radiation (brachytherapy), the radiation comes from radioactive material that is sealed in needles, seeds, wires, or catheters and placed directly in or near the tumor. Radiation therapy is a very important tool in the treatment of cancer. More than half a million cancer patients receive radiation therapy each year, either alone or in conjunction with surgery, chemotherapy, or other forms of cancer therapy. Radiation therapy is useful in cases where surgical removal of the cancer is not possible or when surgery might weaken the anatomy, for example, when tumors are located close to the spinal cord or inside the lung. Together with image-guided treatment planning, radiation therapy is a powerful tool in the treatment of cancer, particularly when the cancer is detected at an early stage. It is estimated that 50% of all cancer patients undergo radiation as part of their treatment. Radiation therapy can be used following surgery to destroy any cancer cells that were not removed by surgery, or prior to surgery to "shrink" a previously inoperable tumor to a manageable size to enable surgical excision. Radiation can also be used to destroy any remaining cancer cells after surgery. Chemotherapy and radiation therapy may also be used together to effectively treat the cancer. Radiation can also be used to help relieve symptoms of advanced cancer (such as bleeding or pain), even if a cure is not possible. Intensity-modulated radiation therapy (IMRT) and Computed Tomography (CT) Imaging are the most widely used techniques. This article presents results of an ongoing research in the development a three-dimensional (3D) prediction model for tumor deformation during radiation treatment for cancer patients. The rationale for this project was to successfully implement a framework for treatment planning using predictive models and to increase the accuracy of the treatments. This remaining portion of this chapter is divided into several sections. In Section 6.2, a comprehensive survey of methods used for tumor modeling is discussed. The problem statement is presented in Section 6.3, which illustrates the significance of the proposed problems. The implemented methodology is presented in Section 6.4. The summary and conclusion with some recommendations for future research are discussed in Section 6.5.

6.2 Background

Modeling and predicting deformations of body organs has been attracting interest in recent years. These anatomical deformations can be caused by diverse factors such as breathing, skull opening, tumor growth, and so on (Kyriacou et al. 2001). Tumor growth is an anatomical deformation and has critical applications in cancer treatment. The mechanisms underlying the expansion of a tumor are very complex and are different for different kinds of tumor (Wasserman and Acharya 1996). The first step in the modeling of tumor growth is the task of detecting the position of the tumor, that is, to recognize and identify gross tumor volume (GTV) in the body of the patient. Usually, a doctor recognizes the GTV and designs its borderlines manually on the computer tomography slice (Zizzari et al. 2001). From the modeling point of view, the first step is to acquire the volumetric visualization of the tumor structure. Volume visualization is concerned with the abstraction, interpretation, rendering, and manipulation of the structure under consideration. Since volumetric objects often come from sampled volume datasets generated with biomedical data acquisition devices such as computed tomography (CT), magnetic resonance imaging (MRI), and confocal microscopy, volume visualization is playing an increasingly important role in biomedical research and applications. Traditionally, however, volume visualization techniques have been used mostly to view the internal structures of volume data. The modeling-rendering paradigm adopted by many computer graphics applications, such as computer-aided design and computer animation, relied on a separate object modeling scheme (i.e., surface and solid modeling) to support object representation, construction, and manipulation. The recent model-based volume visualization approach extends such modeling-rendering paradigms to a volumetric environment where volumetric modeling and rendering are integrated into one model-based visualization system (Pham et al. 2001; Heramb et al. 2004). This integration may also combine geometric and volumetric objects into one intermixed environment without converting one into another. Volume Visualization can be achieved by using image processing techniques. Images may be acquired in the continuous domain such as on X-ray film, or in discrete space as in MRI. In 2D discrete images, the location of each measurement is called *a pixel* and in 3D images, it is called a *voxel*. The Image data are processed using one or a combination of different segmentation techniques available. Image segmentation is defined as the partitioning of an image into nonoverlapping, constituent regions that are homogeneous with respect to some characteristic such as intensity or texture.

Medical images are of great importance in radiation therapy, which became a privileged application field for image processing techniques. Malandain et al. (1999) have discussed the different image processing techniques used for radiation therapy and also list the image matching and image segmentation techniques. These image segmentation algorithms play a vital role in numerous biomedical imaging applications such as the quantification of tissue volumes (Lane and Abukmeil 1998). Worth et al. (1997) proposed the use of segmentation techniques in the study of anatomical structures. They used the image segmentation techniques to study the abnormality in the cerebral structure in schizophrenia. Segmentation studies proposed by Worth et al. (1997) suggest that gray matter is reduced but that white matter volumes may actually be increased for the anatomies under consideration. Pham et al. (2001) proposed methods for performing segmentations that vary widely depending on the specific application, imaging modality, and other factors. For example, the segmentation of brain tissue has different requirements from the segmentation of the liver. General imaging artifacts such as noise, partial volume effects, and motion can also have significant consequences on the performance of segmentation algorithms. Furthermore, each imaging modality has its own idiosyncrasies with which to contend. There is currently no single segmentation method that yields acceptable results for every medical image. Methods do exist that are more general and can be applied to a variety of data. However, methods that are specialized to particular applications can often achieve better performance by taking into account prior knowledge. Selection of an appropriate approach to a segmentation problem can therefore be a difficult dilemma. In the work presented by Pham et al. (2001), nearly all medical images used for image segmentation were represented as discrete samples on a uniform grid. Segmentation methods typically operate on the same discrete grid as the image. However, certain methods such as deformable models are capable of operating in the continuous spatial domain, thereby providing the potential for subpixel accuracy in delineating structures. Subpixel accuracy is desirable particularly when the resolution of the image is on the same order of magnitude as the structure of interest. Segmentation on the continuous domain is not equivalent to partial volume estimation or other soft segmentation methods. Partial volume estimation methods merely provide the fraction of a structure that is present in a voxel. This may be sufficient for quantification purposes but not in situations where precise localization is required, such as for tumors in surgical or radiotherapy planning. Continuous segmentation methods actually reconstruct how a structure passes through a voxel. Although continuous segmentation methods have subpixel or subvoxel resolution, their precision and accuracy are still dependent on the resolution of the original data. Furthermore, this level of precision can be difficult to validate on real data. Hohne et al. (1989) used a ray casting algorithm working on a gray scale voxel data. They used intensity thresholding for surface rendering purposes. The method was followed determining the surface by imaging the negative distance to the observer. Shen et al. (2001) presented a deformable model for automatically segmenting brain structures from volumetric MR images and obtaining point correspondences, using geometric and statistical information in a hierarchical scheme. The focus of the model was on the most reliable structure of the respective targets and follows a hierarchical scheme. They used this technique to segment boundaries of the ventricles, whereas the models proposed by Cootes et al. (1995) and Wang and Staib (1998) did not use weighted importance to the different shapes but used equal weights when calculating shape statistical parameters. Cootes et al. (1995) used two steps for an active shape model (ASM) algorithm, namely, image data interrogation and shape approximation. They considered the shape approximation in detail, and the model can be considered the first statistical shape model for medical image analysis. Their algorithm considered the 2D problem in which a modification of the error term permits a closed-form approximate solution that then can be used to produce starting estimates for the iterative solution. Wang and Staib (1998) from Yale University used elastic and statistical shape models for nonrigid registration. The formulation of their model was based on the rule that force is directly proportional to the displacement. Ferrant et al. (2001) modeled the biomedical shape changes using a physics-based model of the objects the image represents. They used the shape changes of the surfaces of the objects as boundary conditions to the physics-based FE model. It allows inferring a volumetric deformation field from the surface deformations. The shape information of the objects in the image sequence was extracted using an active surface model, and the changes the objects undergo were characterized using a physics-based model. The method followed was to track the boundary surfaces in the image sequence, and use the boundary motion as input for the FE model. This boundary motion was used as a boundary condition for the FE model to infer a volumetric deformation field. In the algorithm, the labeled 3D image from which the mesh needs to be computed is first divided into cubes of a given size, which are further divided into five tetrahedra with an alternating pattern so as to avoid diagonal crossings on the shared quadrilateral faces of neighboring cubes. The initial cube size determines the size of the largest tetrahedra the mesh will contain. Each tetrahedron is checked for subdivision according to the underlying image content. Ferrant et al. (2001) decided to only subdivide tetrahedra that lie across boundaries of given objects, so as to obtain a detailed description of their boundaries. The edges of those tetrahedra to be subdivided are labeled for subdivision, and a new vertex is inserted at their middle point. This process is executed iteratively until the smallest edges have reached a specified minimum size. The assumption in this methodology was that the objects that are being imaged have an elastic behavior during deformation. Miller (1998) and Miller and Chinzei (1997) proposed a constitutive model for brain tissue for modeling and simulation of surgical procedures. The model used the finite element method based on the strain energy function in polynomial form with time-dependent coefficient. The model considered and combined geometrical nonlinear ties and time-dependent nonlinearities. The model lacked effectiveness in its finite element implementation as the model was nonlinearly viscoelastic. Miller (1998) provided an alternative approach with the use of the hyperelastic, linear viscoelastic constitutive model. This was implemented in finite element packages such as ABAQUS. The ability of a linear viscolelastic model to describe the deformation behavior is not as perfect as that of a fully nonlinear model. However, the author proposed that simplicity of the finite element code outweighs the slight loss in accuracy.

Mohamed et al. (2001) presented a statistical approach for predicting deformations induced in the brain anatomy due to tumor growth. The approach utilizes the principal modes of covariation between deformed and undeformed anatomy to estimate the one given the other, that is, with a statistical model constructed from a number of training samples, a patient brain anatomy prior to tumor growth is estimated based on the patient's tumor-bearing images. This proposed approach was tested on a data set of 40 axial 2D brain images of normal human subjects. The approach used in the model was applied using the statistical framework proposed by Shen et al. (2001); Shen et al. (2001) proposed a framework for modeling and predicting anatomical deformations and tested on simulated images. The framework can be used for modeling a variety of deformations. The authors emphasized surgical planning, and particularly on modeling and predicting changes of anatomy between preoperative and intraoperative positions, as well as on deformations induced by tumor growth. Two methods are examined in this framework. The first is purely shape-based and utilizes the principal modes of covariation between anatomy and deformation in order to statistically represent deformability. When a patient's anatomy is available, it is used in conjunction with the statistical model to predict the way in which the anatomy will/can deform. The second method is related, and it uses the statistical model in conjunction with a biomechanical model of anatomical deformation. It examines the principal modes of covariation between shape and forces, with the latter driving the biomechanical model, and thus predicting deformation. Results are shown on simulated images, demonstrating that systematic deformations, such as those resulting from change in position or from tumor growth, can be estimated very well using these models. Estimation accuracy will depend on the application, and particularly on how systematic a deformation of interest is.

6.3 Problem Significance

Head and neck cancer is the name given to a variety of malignant tumors that occur in the head and neck region. By definition, this region usually excludes tumors that occur within the brain; those tumors usually are treated by neurosurgeons and associated specialties. Tumors from other parts of the body can spread to the head and neck region as well. The most common type of malignant tumor in the head and neck region is squamous cell cancer, also known as squamous cell carcinoma. The lining of much of the mouth, nose and throat is made up of a type of cell known as squamous cell. When a malignancy arises in these cells, the tumor is called squamous cell carcinoma. Treatment for head and neck cancer is very much dependent on the type of tumor, where it has occurred, and how large the tumor is. The goal in all cancer treatment is to remove the tumor with as little damage as possible to important structures in the head and neck. The three main types of treatment for managing head and neck cancer are radiation therapy, surgery, and chemotherapy.

The primary treatments are radiation therapy or surgery, or both combined. Tumor geometry changes over time with or without a treatment. Radiation therapy is used as a noninvasive surgery alternative for treating cancer patients. Radiation hinders cancerous cells from growing; hence, the tumor volume is likely to stop its growth and shrink its size. Because of the radiation dose limit that a human tissue can tolerate without a substantial side effect, 2 Gy of radiation is delivered to the patient daily. Since most types of tumors require higher than 60 Gy for the treatment, a radiation treatment will take place over the period of 5 to 7 weeks. A master plan for the treatment is developed at the beginning of the patient's visit to the hospital. Then, 2 Gy of radiation is delivered to the same tumor volume as seen in Day 1. *This is a critical flaw.* Some researchers modeled deformation of human organs using biomechanical models. These models using biomechanical properties are considered to be accurate as they utilize the physical knowledge about the deformed anatomy and its properties, assuming that the material properties, equations governing the deformations, and the boundary conditions are known (Andersen et al. 2000; Asachenkov et al. 1989; Garbey and Zouridakis 2003). However, these factors are not always known, and they are very complex to determine.

In the proposed planning system, we will monitor and predict daily tumor volume changes during the entire treatment period. Finally, the treatment plan will adapt to the changes in the tumor volume. The proposed methodology aims at increasing the accuracy of each therapy. Figure 6.1 illustrates the scope of the proposed model.

6.4 Proposed Methodology and Supporting Technology

Many types of cancer have been treated by radiation therapy in recent years. The treatment objective is to achieve tumor control by planning a significant total dose of radiation to the cancerous region to sterilize the tumor without damaging the surrounding healthy tissues. Although the size and shape of the tumor may shrink due to the radiation or may even expand for some other unknown reasons, there

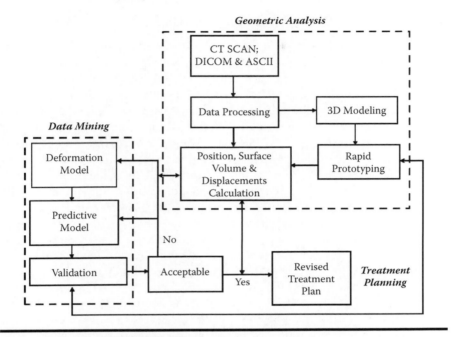

Figure 6.1 Scope of the proposed methodology.

is a critical need to understand how tumor geometry changes over time during the radiation treatment. The objective of this project was to develop a three-dimensional (3D) prediction model for tumor deformation during radiation treatment for cancer patients. Most advanced geometric and surface modeling techniques will be used rather than repetitive CT (computer tomography), MRI (magnetic resonance imaging) scans, or biomedical models. Clinical patients were obtained from the University of Texas–MD Anderson Cancer Center (MDACC). Our rationale for this project was that its successful completion is expected to provide a framework for treatment planning using predictive models that will significantly increase the accuracy of the treatments and quality of life for patients with head and neck tumors. The objective is to develop a mathematical model that captures the changes in the tumor volume. No literature has addressed mathematical models to quantify the tumor volume change such as the shape and the location of the tumor geometry. A mathematical model will be developed to evaluate the change in the tumor volume and its impact on treatment plans. Our 3D deformation model is based on CT image data. The proposed project is implemented in three major phases:

- Data collection and analysis
- Rapid prototyping model development
- Geometric modeling of deformations

6.4.1 Phase I—Data Collection and Analysis

Recent advances in image-based treatment planning have improved the clinician's ability to design conformal treatment plans that maximize both tumor coverage and normal tissue sparing (Chao et al. 2000; Chao et al. 2003; Dawson et al. 2000; Eisbruch et al. 1999; Eisbruch et al. 2001). Patients receiving RT to the head and neck will experience significant anatomic changes during their treatment course, including shrinking of the primary tumors and changes in overall body habitus/weight loss (Suit and Walker 1980; Barkley and Fletcher 1977; Sobel et al. 1976 and Trott 1983). Several variables could cause deviations in radiation dose delivery from the initial treatment plan. Aside from variations in the initial target volume and normal tissue delineation, these uncertainties include external daily setup variations and internal geometric and volumetric changes occurring throughout the 6–7-week RT course. The methods available for quantification of such changes have been rather crude. The studies of radiation effect on the gross tumor volume (GTV) and position, for instance, have been limited to recording the gross disease characteristics by methodical physical examinations before, during, and at the completion of RT (Suit and Walker 1980; Barkley and Fletcher 1977; Sobel et al. 1976; Trott 1983). These studies did not adequately describe the ongoing geometric changes in tumors and normal tissues during a several-week course of treatment. The goals for this phase are to specifically generate the necessary data to quantify the progressive geometric/anatomic changes occurring in patients treated with RT for head and neck cancer. The initial studies have shown some correlation among data (Barker et al. 2004). The data are in 3D ASCII format with x, y, and z coordinates. Figures 6.2 and 6.3 illustrate input data in DICOM format (DICOM format is used for prototyping) and the contours generated using point clouds data (x, y, and z format) in MATLAB® for geometric analysis.

6.4.2 Phase II—Prototyping and Rapid Prototyping Model

The collected data is further processed using MagicRP® software (Figure 6.4). The analysis includes editing the STL (stereolithography) data file and the initial/visual analysis of geometrical and dimensional features of tumors. The result is used to develop prototype mockups of different stages of tumors and their reference to the main axis of the skull structure. The mockup prototypes are built using the Z-Corp 406® color prototyping machine as shown in Figure 6.4. The prototypes are used to visually inspect the position of the tumors as they go through deformation after each therapy with reference to the main axis of the skull structure. The prototype is used for validation and verification of the model. The prototype models are illustrated in Figures 6.5 and 6.6.

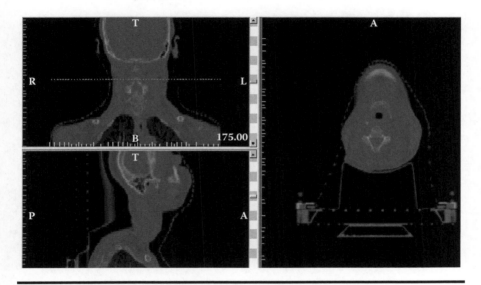

Figure 6.2   Input data for RP analysis in DICOM format.

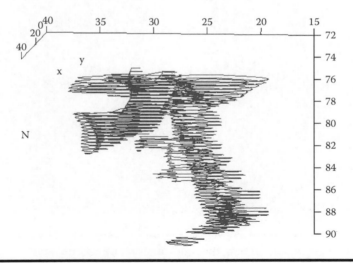

Figure 6.3   Sample output data from MATLAB® (contours from point cloud).

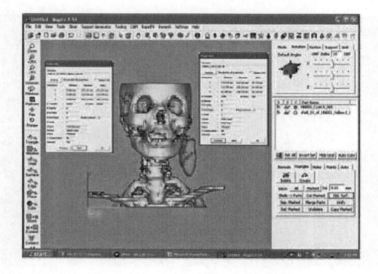

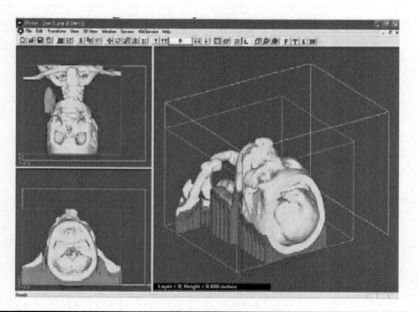

Figure 6.4   PLY file for building the mockup using RP technology.

Figure 6.5 Prototyped model for validation.

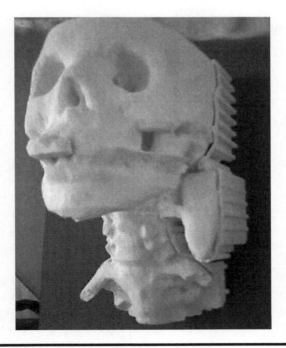

Figure 6.6 Tumor location with reference to the skull.

6.4.3 Phase III—Geometric Modeling and Analysis

Steps in this phase include

■ Surface contouring and construction
■ Volume construction and analysis

The data is in 3D ASCII format with x, y, and z coordinates. There are two major tasks associated with this phase of the tumor deformation geometry modeling:

■ To visualize the tumor contour and map the surface of the tumor volume.
■ To establish tumor's three selected features: center coordinate, surface area, and volume.

The processed data include the ASCII data that are sorted in z coordinates. Points with the same z value (same layer) are then used to construct the slice contours. These generated contours are connected together to build the 3D surface. Both the triangular patch and the rectangular patch approaches are used to create the connected contours. Both methods are formulated. The primary objective is to illustrate the surface geometry with the best level of accuracy. For building these patches, r is defined as the Euclidean distance between two points m and m' where m and m' are points on two different layers. This is shown below for all (x,y,z) in tumor volume (6.1)

$$r = \sqrt{\left(m_x - m'_x\right)^2 + \left(m_y - m'_y\right)^2 + \left(m_z - m'_z\right)^2} \tag{6.1}$$

r_{min} is the $min(r_i)$ for i from 1 to *num*, where r_i is the distance from m to m_i'; m_i' is the point on the $n+1$-th layer, and *num* is the number of points in the $n+1$-th layer. Using the four-point approach, two points m and point $m+1$ on the n-th layer are selected. Using the r_{min} definition, m's nearest neighbor m' on the $n+1$-th layer is determined. The same analysis is used for all other points. m initially is a random point selected on the nth layer, and all other points on the nth layer are searched using the nearest neighbor method. Finally, all m_{nums} patches are built. For the triangle method, on the nth layer, two initial points for the first triangle are selected. Using the same r formula, the nearest neighbor point for the first point is determined, and these three points are used to define the triangle patch. The same process continues until all points are connected into a patch. Figures 6.7 and 6.8 illustrate constructed surfaces between layers using both techniques. For our analysis, the triangular method is used owing to its simplicity and software performance, although future research may be necessary to identify which method is more accurate. In order to calculate the changes in the tumor location, the center of the tumor is calculated using the average values of all coordinates (T1) and a reference point

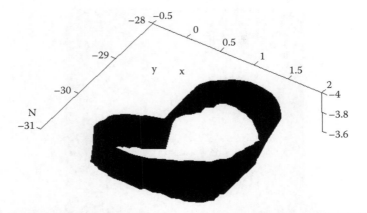

Figure 6.7   Surface using four-point patches.

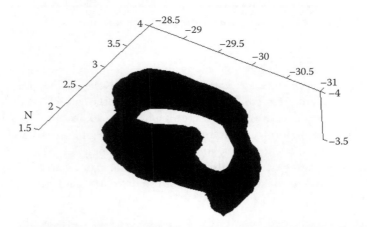

Figure 6.8   Surface using three-point patches.

on the spine canal (C2_bone). T2 (Figure 6.9) illustrates a sample of CT data and the processed 3D image of the bone structure near the head and neck cancer tumor.

The following expression is used to calculate the displacement in tumor position for all (x,y,z) in tumor volume (6.2):

$$Dis(T1,T2) = \sqrt{(x_{T1} - x_{T2})^2 + (y_{T1} - y_{T2})^2 + (Z_{T1} - Z_{T2})^2} \qquad (6.2)$$

To calculate the volume of the tumor, the extreme points are used to create the required rectangles. These rectangles are girded using reference cubes. The total

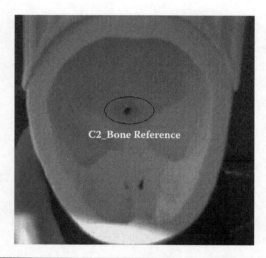

Figure 6.9 View of c2-Bone area on the prototyped skull.

number of cubes multiplied by the volume of the reference cube is used to estimate the overall volume. A more accurate method will be developed for this project and tested using the fabricated prototypes. The volumes for the features on the edges of the tumor are calculated using intersection points. A series of MATLAB® routines are developed to perform the required calculations. Figure 6.10 illustrates the result of surface contouring and volume constructions for a sample tumor after 15 radiation therapy sessions. The analysis result is illustrated in Figures 6.11 and 6.12. Tables 6.1 and 6.2 summarize the results of tumor volume and surface values on average for all the patients.

6.5 Conclusion

Many cancer tumors are treated with radiation therapy these days. The objective is to deliver the correct amount of radiation accurately to the tumor and prevent the surrounding normal tissues from being damaged by radiation. A master treatment plan is developed based on pretreatment CT scans at the time of the patient's first visit to the hospital. Most treatment planning systems assume that the tumor geometry will not change over the entire period. There is a critical need to monitor the changes in the cancer tumor during the radiation therapy period to understand the changes of tumor shape and size in order to prevent significant effects of inaccuracy in planning. The proposed methodology provides a framework that can be used for analyzing and predicting tumor growth for patients with head and neck tumors.

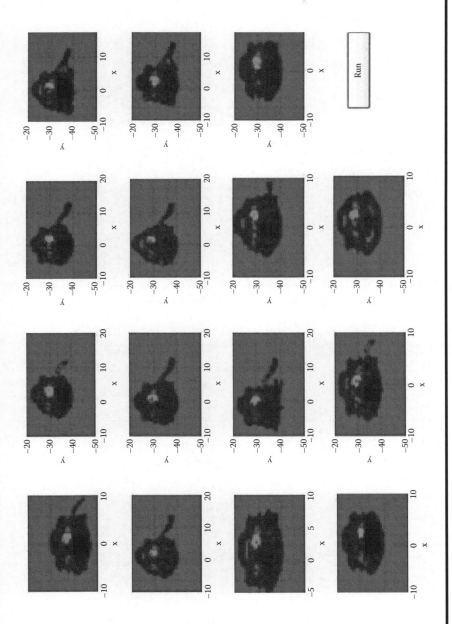

Figure 6.10 Sample of geometry reconstruction using MATLAB®.

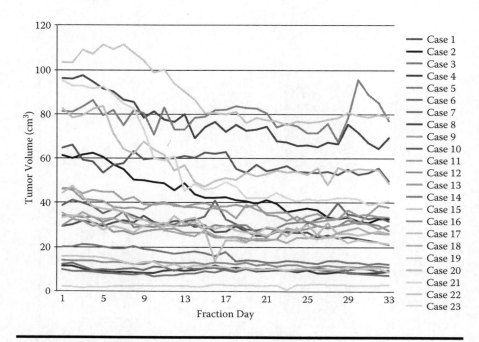

Figure 6.11   Tumor volume responses to radiation therapy.

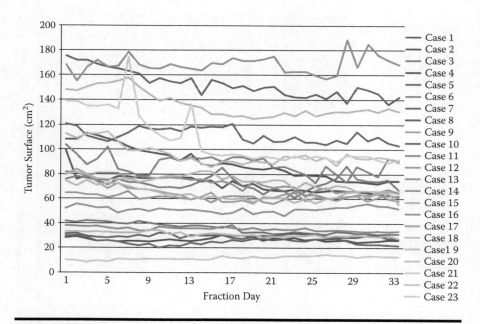

Figure 6.12   Tumor surface responses to radiation therapy.

Table 6.1   Average Tumor Volume Values

							Volume (cm³)					
Patient #	1	2	3	4	5	6	7	8	9	10	11	12
Average	27.86	44.18	16.13	9.40	8.82	33.15	10.16	76.78	37.88	57.42	37.52	30.28
Patient #	13	14	15	16	17	18	19	20	21	22	23	
Average	27.66	79.30	88.64	12.39	29.99	27.46	11.74	59.3	56.98	27.75	2.54	

Table 6.2   Average Tumor Surface Values

						Surface (cm^2)						
Patient #	1	2	3	4	5	6	7	8	9	10	11	12
Average	72.99	89.07	35.83	26.80	25.05	72.8	29.65	152.7	72.42	111.5	87.82	61.56
Patient #	13	14	15	16	17	18	19	20	21	22	23	
Average	51.4	168.45	136.22	35.35	64.51	63.83	30.95	96.06	108.2	74.1	11.8	

Authors

Maryam Azimi is a Software Project Manager at Lenovo Corporation. She received her PhD in industrial engineering from the University of Houston in 2011. Her research interests are systems engineering, data mining in health care, lean Six-Sigma, and project management.

Ali K. Kamrani is an Associate Professor of Industrial Engineering. He is also Founding Director of the Design and Free Form Fabrication Laboratory at the University of Houston, USA. He received his BS in Electrical Engineering in 1984, Master of Engineering in Electrical Engineering in 1985, Master of Engineering in Computer Science and Engineering Mathematics in 1987, and PhD in Industrial Engineering in 1991, all from the University of Louisville, Louisville, Kentucky. His research has been motivated by the fundamental application of systems engineering and its application in advanced design and development of complex systems. He is the Editor-in-Chief for the *International Journal of Collaborative Enterprise* and the *International Journal of Rapid Manufacturing*.

Emad Samir Abdelghany is an Associate Professor in Industrial Engineering Department, College of Engineering, King Saud University, and Assistant Professor in Mechanical Engineering Department at Helwan University, Faculty of Engineering, Helwan, Cairo, Egypt. He received his PhD in Industrial Engineering from the University of Houston, TX, USA in 2005. His current research focuses on CAD, CAM, rapid prototyping, advanced manufacturing systems, and collaborative engineering.

References

Andersen, P.R., Bookstein, F.L., Conradsen, K., Ersboll, B.K., Marsh, J.L., and S. Kreiborg. 2000. Surface bounded growth modeling applied to human mandibles. *IEEE Transactions on Medical Imaging* 19(11): 1053–63.

Asachenkov, A.A. 1989. Nonlinear models and data analysis of cancer patients. *IEEE International Symposium on Circuits and Systems* 3: 1617–19.

Barkley, H.T, and G.H. Fletcher 1977. The significance of residual disease after external irradiation of squamous-cell carcinoma of the oropharynx. *Radiology* 124: 493–95.

Barker, J., Garden, A.S., Ang, K.K., O'Daniel, J.C., Wang, H., Court, L.E., Morrison, W.H., Rosenthal, D.I., Chao, K.S., Tucker, S.L., Mohan, R., and Dong, L. 2004. Quantification of volumetric and geometric changes occurring during fractionated radiotherapy for head-and-neck cancer using an integrated CT/linear accelerator systems. *International Journal of Radiation Oncology Biology Physics* 59(4): 960–70.

Chao, K.S., Low, D.A., Perez, C.A., and Purdy, J.A. 2000. Intensity-modulated radiation therapy in head and neck cancers: The Mallinckrodt experience. *International Journal of Cancer* 90: 92–103.

Chao, K.S., Ozyigit, G., Tran, B.N., Cengiz, M., Dempsey, J.F., and Low, D.A. 2003. Patterns of failure in patients receiving definitive and postoperative IMRT for head-and-neck cancer. *International Journal of Radiation Oncology Biology Physics* 55: 312–21.

Cootes, T.F., Cooper, D., Taylor, C.J., and Grahant, J. 1995. Active shape models—their training and application. *Computer Vision and Image Understanding* 61(1): 38–59.

Dawson, L.A., Anzai, Y., Marsh, L., Martel M.K., Paulino, A., Ship, J.A., and Eisbruch, A. 2000. Patterns of local-regional recurrence following parotid-sparing conformal and segmental intensity-modulated radiotherapy for head and neck cancer. *International Journal of Radiation Oncology Biology Physics* 46: 1117–26.

Eisbruch, A., Ship, J.A., Kim, H.M., and Ten Haken, R.K. 2001. Partial irradiation of the parotid gland. *Seminars in Radiation Oncology* 11: 234–39.

Eisbruch, A., Ten Haken, R.K., Kim, H.M., Marsh, L.H., and Ship, J.A.. 1999. Dose, volume, and function relationships in parotid salivary glands following conformal and intensity-modulated irradiation of head and neck cancer. *International Journal of Radiation Oncology Biology Physics* 45: 577–87.

Ferrant, M., Macq, B., Nabavi, A., and S.K. Warfield. 2001. Defonnable Modeling for Characterizing Biomedical Shape Changes. Surgical Planning Laboratory, Brigham and Women's Hospital, Harvard Medical School.

Garbey, M., and G. Zouridakis. 2003. Modeling tumor growth: From differential deformable models to growth prediction of tumors detected in PET images. *Proceedings of the 25th Annual International Conference of the IEEE EMBS* 2687–90.

Heramb, I., Kamrani, A., and R. George. 2004. Rapid prototyping applications in cancer tumor deformation modeling. *34th International Conference on Computers and Industrial Engineering*, November 14–16, San Francisco, CA.

Hohne, K.H., Bomans, M., Pommert, A., Riemer, M., Schiers, C., Tiede U., and G. Wiebecke. 1989. 3D-Visualization of tomogaphic volume data using the generalized voxelmodel. Chapel Hill Volume visualization Workshop. 51–57.

Kyriacou, K.S., Davatzikos, C., and A. Mohamed. 2001. A statistical approach for estimating brain tumor induced deformation. *IEEE Workshop on Mathematical Methods in Biomedical Image Analysis*. 52–59.

Lane, S.M., and S.S. Abukmeil. 1998. Brain abnormality in schizophrenia: a systematic and quantitative review of volumetric magnetic resonance imaging studies. *British Journal of Psychiatry* 172: 110–120.

Malandain, O., Bondiau, P.Y., Chanalet, S., Marcy, P.Y., Foa, C., and Ayache, N. 1999. Medical Imaging Applications. Technical Note.

Miller, K. 1998. Finite Deformation, Linear and Non-Linear Viscoelastic Models of Brain Tissue Mechanical Properties. Technical Note.

Miller K., and K. Chinzei. 1997. Constitutive modeling of brain tissue, experiment and theory. *Journal of Biomechanics* 30(11/12): 1115–21.

Mohamed, A., Kyriacou, S.K., and C. Davatzikos. 2001. A statistical approach for estimating brain tumor induced deformation. IEEE Workshop on Mathematical Methods in Biomedical Image Analysis, 52–59.

Pham, L.D., Chenyang, X. Yezzi, A., and Prince J., 2001. A Summary of Geometric Level-Set Analogues for a General Class of Parametric Active Contour and Surface Models. Proceeding of 2001 IEEE Workshop on Variational and Level Set Methods in Computer Vision (VLSM 2001), 104–111.

Shen, D., Herskovits, E.H., and C. Davatzikos. 2001. An adaptive-focus statistical shape model for segmentation and shape modeling of 3-D brain structures. *IEEE Transactions on Medical Imaging* 20(4): 257–70.

Sobel, S., Rubin, P., Keller, B., and Poulter, C. 1976. Tumor persistence as a predictor of outcome after radiation therapy of head and neck cancers. *International Journal of Radiation Oncology Biology Physics* 1: 873–80.

Suit, H.D., and Walker, A.M. 1980. Assessment of the response of tumors to radiation: Clinical and experimental studies. *British Journal of Cancer Suppl* 41: 1–10.

Trott, K.R. 1983. Human tumor radiobiology: Clinical data. *Strahlentherapie* 159: 393–97.

Wang, Y., and Staib, L.H. 1998. Elastic model based non-rigid registration incorporating statistical shape Information. *MICCAI* 1162–73.

Wasserman, R., and Acharya, R. 1996. A patient specific in vivo tumor model. *Mathematical Biosciences* 13: 111–40.

Worth, A.J., N. Makris, Caviness, V.S., and Kennedy, D.N. 1997. Neuroanatomical segmentation in MEl: technological objectives. *International Journal of Pattern Recognition and Artificial Intelligence* 11:1161–87.

Zizzari, A., Seiffert, U., Michaelis, B., Gademann, G., and Swiderski, S. 2001. Detection of tumor in digital images of the brain. *Proceedings of the LASTED International Conference on Signal Processing, Pattern Recognition and Applications* 132–37.

Chapter 7

Product Variety and Manufacturing Complexity

Ali K. Kamrani, PhD, PE

Industrial Engineering Department, University of Houston, Houston, Texas

Contents

Product variety is defined as the total number of products, product variants, or features that a manufacturer offers to its customer. The designers must focus on the important features based on the customer requirements. For example, Quality Function Deployment (QFD) is used to map a hierarchical structure that will identify the product features according to the importance to the customer. Manufacturers should consider all possible features while designing a new product or developing product variants. They must also consider the impact on their manufacturing operations. From the engineering point of view, engineers should design the product lines in such a way that the manufacturing cost is minimized. This includes all varieties and the base product line. For example, the product structure graph could assist designers and engineers to better design the varieties and the flexible production line to produce products effectively (Martin and Ishii 1996). This allows the designer to identify which components are standard and which subassemblies are to be modularized.

New features are typically added to base product line due to sales, marketing, or new customer requirements. The variety of decisions is classified into strategic or tactical (Martin and Ishii 1996). Strategic variety decisions are represented by the number of products that are being offered to the customer. Typically, the product manager will decide on what types of products to offer based on the new customer requirements and technological advancements. These decisions are made based on trade-offs between the benefits and costs of adding new varieties. Tactical variety decisions are made at the design engineer level. These are decisions that mainly affect how the products and their variants are produced. These decisions require integrations between both the design and manufacturing departments. Product variety is further classified into internal and external varieties. From these, the external variety is the one that is seen by the customers. The external variety is divided into (1) value addition for the customer; desired by customer and (2) no value addition for the customer; confuse the customer. Manufacturers believe the larger the variety of products, the higher the profit, although excessive product variety is a major problem since it adds cost and complexity to the system (Christiansen 2003). Therefore, manufacturers must eliminate the unnecessary variants and focus on the variants based on the customer needs and technological advancements. A demand for variety exists only due to the heterogeneous nature of consumers and their individual preferences. Individual consumer typically seeks variety in their own point of view. The degree of variety from the consumer's point of view is measured as (Tang and Yam 1996)

$$\sum_{i=1}^{n} M_i$$

where
 M_i = Number of different models within brand i
 n = Total number of different brands available.

7.1 Design for Variety

Design for variety (DFV) is a new methodology used for measuring the impact of variety (Martin and Ishii 1996). The DFV methodology is used to design the production system in order to satisfy the customer needs in a timely manner with minimum cost ands maximum quality at minimum product life cycle. DFV is not designed for minimum cost since minimum cost is easily achieved by eliminating any variety. Market changes (global market, global economy, and new innovations) and international competition are forcing many manufacturers to consider increasing and producing product varieties. Most of these companies must also try to reduce the time to market. This will reduce the high cost of design and development. For example, Toyota reduced the development time for its automobiles from 24 months to 18 months by 1998. The new car development program in the United States ranges between 3 to 4 years, while in Japan it is less than 3 years (Prasad 1997). The key strategy in the DFV methodology is to identify the standard model and to utilize the developed strategy to design products and their processes that lead to process time reduction, lowering inventory, and lowering logistics costs. The DFV methodology is a procedure that helps managers and engineers to better understand the true costs associated with introducing new product variants into their product line. This is illustrated by using cost indices. The cost of variety is reduced by (Martin and Ishii 1996)

- Differentiate as late as possible
- Shorten the time between the processes
- Reduce setup costs
- Delay the addition of value to the product to later in the process flow.

DFV is also based on the concept of concurrent engineering and integrated approach to design and manufacturing. From the engineer's point of view, DFV extends design for manufacturing (DFM) to include marketing service, and other product life cycle issues that could impact cost and profit. It is concluded that variety and introduction of product variant should be handled as close to the customer demand as possible. Most of the literature on product variety has focused on (Da Silveira 1998)

- Its importance within the competitive strategy
- Its impact on operations performance
- The use of flexibility for dealing with product and their variety in operations strategy

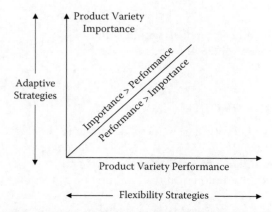

Figure 7.1 The role of adaptive and flexibility strategies in product variety management.

Product variety management focuses on (1) importance or (2) performance. For the importance issue, companies may limit either the product variety or the objective associated with variety. This is done by using priorities setting, focused manufacturing, or mass customization. For the performance issue, companies may increase the variety performance by improving flexibility thorough setup reduction, cellular manufacturing, design for manufacturability, and flexible technologies. Figure 7.1 represents the differences between product variety importance and the product variety performance (Da Silveira 1998). The four elements of product variety management include

- The strategic importance of product variety
- The impact of increasing variety in performance
- The management of product and part variety through adaptive flexibility strategies
- The flexibility types related to flexibility strategies

7.2 Costs of Variety

Increasing product variety will increases the number of features per part and complexity in part design and its manufacturing. Therefore, the costs due to these complexities will increase, which requires careful analysis. The question is how to find best the level of variety that could result in the benefits of volume and variety. The product architecture is measured by the number of functions (features) per components in the product. For example, a modular architecture has a low number of functions per component, and integral architecture has a high number of functions per component (Ishii et al. 1997).

A broader product line with corresponding low volumes for each item could result in higher unit costs, mainly because of increases in overhead expenses. Also, by choosing a modular architecture, the designer may sacrifice some performance, but could gain the benefits of a much more robust design for maintainability, manufacturability, and serviceability. Modularity can be accomplished in different ways as described in previous chapters. The designer must also evaluate the direct and indirect costs of adding variety. The direct costs of adding variety are (Martin and Ishii 1996)

- Capital equipment
- Personnel training
- Engineering time to make new drawings, analyze the new design, run qualification tests, etc.
- Adding supplies

The indirect costs (overhead) of adding variety are (Martin and Ishii 1996)

- Raw material inventory
- Work-in-process (WIP) inventory
- Finished goods inventory
- Postsales service inventory
- Reduction in capacity due to setups
- Increased logistics of managing variety

The indirect costs are always difficult to determine and, in many cases, estimated values are used for analysis.

The variety cost is directly impacted by the number of part's functions and features. When fewer functions and features are incorporated, studies have shown the complexity will be in the form of assembly cost. On the other hand, when a large number of features and functions are incorporated per part, the number of parts produced will be small, but due to the design complexity, part cost will be high. A proposed measure of cost of variety is (Prasad 1997; Martin and Ishii 1996):

$$\text{Cost of variety} = \text{min cost of manufacturing an assembly} * (1 - C_v) \\ + \text{max cost of manufacturing an assembly} * (C_v)$$

where

$$C_v = \prod_{i=1}^{i=3} (\alpha i)$$

$0 \le \alpha_i \le 1$

$i = 1,2,3$

α_1 = Number of variation in the product

α_2 = Time measured to the finish stage

α_3 = Changeover effort

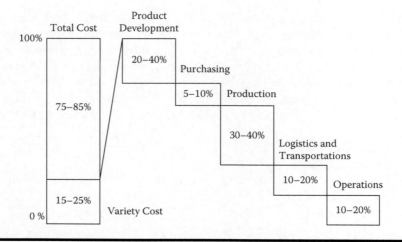

Figure 7.2 Impact of variety on total cost.

These parameters are user defined. The impact of design variety on the total cost is shown in Figure 7.2 (Christiansen 2003).

Manufacturing cost can be classified based on the product variety and the volume of production. Manufacturing costs will decrease when the total volume increases: mass production. Variety-related costs include setup, material handling, inventory, and other overhead costs. In this case, as the variety of products increases, the complexity increases since increasing product variety has a direct impact on the required number of processes (job shop). This results in increasing costs of manufacturing. The objective is to find the best volume versus variety combination that will results in lower overall cost. Cellular manufacturing and flexible manufacturing are the best fit for increasing the level of variety, while reducing the cost of their production. Figure 7.3 illustrates a framework of possible direct and indirect impacts on manufacturing based on product variety (Yeb and Chu 1991).

7.3 Qualitative Methods for Managing Product Variety

7.3.1 Product Structure Graph

The product structure graph (PSG) illustrates the product variety in a hierarchical tree and allows engineers to focus on critical features. This will allow them to identify and eliminate unnecessary varieties. It also assists in determining which components are considered as standardized and which subassemblies are to be modularized. The inputs to the PSG should be the scope of components and features within the product line. For example, the inputs could be from QFD analysis (Martin and Ishii 1996) or include product permutation and the combination of these permutations that creates the different product variants. The output of PSG

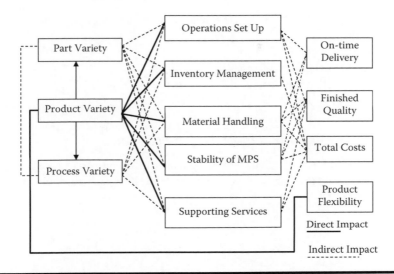

Figure 7.3 Impact of product variety on the manufacturing system. (From Yeb, K., and C. Chu. 1991. *International Journal of Operations and Production Management* **11(8): 35–47.)**

is the product structure, including variants. Designers could use this information to better design and achieve optimum design combinations. The major goal of using the product structure graph is to optimize the manufacturing process used for producing the variety in the given product line, while minimizing the investment required. The graph provides only a qualitative guide to the design of varieties. The product structure graph and complexity measures were applied to a variety of problems, including automotive window regulator, heat trace cable connectors, and hard disk drives (Martin and Ishii 1996). The graph and the complexity measures will clarify the overall product structure, identify cost drivers, and provide a visual guide to redesign opportunity.

7.3.2 Process Sequence Graph

The process sequence graph illustrates the flow of the process sequence and its differentiation points. Differentiating the product later in the assembly process will reduce inventory costs and the complexity of the manufacturing system (Lee and Billington 1993). These strategies depend on the manufacturing time with respect to the required lead time (Martin and Ishii 1997).

7.3.3 Commonality Graph Method

Another method used to manage the variety and its cost is the application of the commonality graph method. In this method, a series of charts are developed based on the selected industry that correlates the commonality of components to (Martin and Ishii 1997)

- Process sequence
- Lead time of components
- Amount of variety desired by customer

7.3.3.1 Process Sequence versus Commonality

The relationships between the commonality of features and the process sequence are presented using this graph. A commonality index for each component (CI_{comp}) is calculated as

$$CI_{comp} = 1 - \frac{U-1}{V_n-1} \qquad (7.1)$$

where
U = number of unique part numbers
V_n = final number of varieties offered

If only one component is considered sufficient for all the required varieties, the CI_{comp} is set to 1. This is considered the desired options.

7.3.3.2 Lead Time versus Commonality

Standardization can be eliminated if there is a low level of commonality and short lead time among the components. If there is a long lead time, 100% commonality is desired in order to minimize the inventory costs associated with the safety stock level and to establish standardized components.

7.3.3.3 Customer Requirements versus Commonality

Variety voice of the customer (V^2OC) is used as a measure of importance for identifying the component variety demanded by customers. It represents the importance of the component to the customers as well as the heterogeneity of the market. One of the techniques to measure that attribute is through conjoint analysis.

7.3.3.3.1 High Variety Low Volume (HVLV)

The principle of lean manufacturing (LM) is modified to match the high variety and low volume (HVLV) conditions (Jina et al. 1997). This framework is illustrated as shown in Figure 7.4.

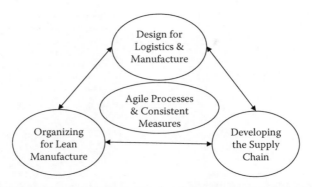

Figure 7.4 Adapting LM principles to HVLV.

7.3.3.4 *Design for Logistics and Manufacture (DFLM)*

DFLM is critical for HVLV since it will reduce the cost and complexity associated with adding variety through

- Common raw material parts
- Common finished parts
- Modular designs
- Staged engineering change control
- Multifunctional teamwork

7.3.3.5 *Organizing for Lean Manufacturing*

Organizing for lean will reduce the variation in the material flow. This could be achieved through organizing the high-level demand (assembly and subassembly) and integrating the customer demand stage and the order release stage.

7.3.3.6 *Integrative Supplier Relationships*

This method proposes the use of generic raw material design, part, and subassemblies from single source rather than multiple vendors.

7.3.3.7 *Process Orientation and Consistent Performance Measures*

This measure is used to monitor the operation's progress. For example, the measures of HVLV situation includes batch sizes, space utilization, setup times, numbers and the justifications for unplanned engineering changes, supplier delivery frequency, customer satisfaction ratings, and delivery time.

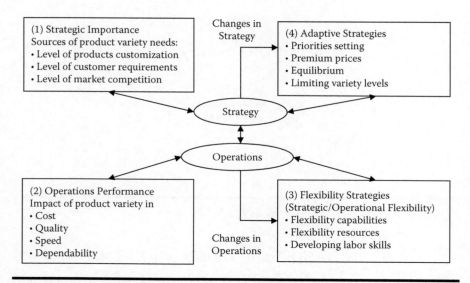

Figure 7.5   A framework of product variety management.

7.3.3.7.1 Empirical Implementation

Empirical approach has also been used to measure the impact and required varieties. One was based on five case studies done in Britain and Brazil by Da Silveira (1998). The primary goal was to develop a product variety management framework. Results from these five case studies (practical results) and literature reviews (theoretical) suggested the framework as illustrated in Figure 7.5.

7.4 Quantitative Methods for Managing Product Variety (Martin and Ishii 1996, 1997)

Measures have been developed that designers could use to measure the impacted costs of adding and providing variety. A commonality index (CI) that represents the utilization of standardized parts is developed. The CI represents the percentage of components used in more than one product model in the manufacturing line. The commonality index is measured as

$$CI = \frac{U}{\sum_{j=1}^{V_n} P_j} \tag{7.2}$$

where

$0 < CI \le 1$

U = number of unique part numbers

P_j = number of parts in model j
V_n = final number of variants offered

A low index number indicates a high level of standardization. If CI is 1, there are no two parts alike. A revised CI_n is also defined based on the number of unique parts.

This is a measure that shows how well the design uses the standardized components. Higher CI_n indicates low usage of unique parts in the design of different varieties. The revised CI_n is represented as

$$CI_n = 1 - \frac{U - \max P_j}{\sum_{j=1}^{V_n} P_j - \max P_j} \qquad (7.3)$$

where
U = number of unique parts
P_j = number of parts in model j
V_n = final number of varieties offered

Another measure that is used is the differential point index (DI). DI considers the process stages that add the variety. In addition, it incorporates the time taken from the differentiation stage to the final process stage. A lower index means that the differentiation is at the later stage. DI is measured as

$$DI = \frac{\sum_{i=1}^{n} d_i V_i a_i}{n d_1 V_n \sum_{i=1}^{n} a_i} \qquad (7.4)$$

where
$0 < DI \le 1$
V_i = number of different products exiting process i
n = number of processes
V_n = final number of varieties offered
d_i = average throughput time from process i to sale
d_1 = average throughput time from beginning of production to sale
a_i = value added to process i

The setup cost index (SI) is another measure that is used. This measure defines the percentage of setups as a function of the total cost. SI is measured as

$$SI = \frac{\sum\limits_{i=1}^{n} V_i C_i}{\sum\limits_{j=1}^{V_n} C_j} \tag{7.5}$$

where

$0 \leq SI \leq 1$

V_i = number of different products in exiting process i

C_i = cost of setup at process i

C_j = total (material, labor, overhead) at the jth product

The cost of variety is estimated using a regression model as

$$\Psi = \beta_0 + \beta_1 CI + \beta_2 DI + \beta_3 SI \tag{7.6}$$

where

Ψ = indirect cost of providing variety

β_0, β_1, β_2, and β_3 are regression coefficients.

7.5 Manufacturing Complexity

One of the biggest challenges of today's industries is the increasing problem of manufacturing complexity due to the variety of products that they have to offer to their customers. From an organizational point of view, complexity is the degree of interaction between the organization's segments that could result in unpredictable behaviors. It is important to understated and determine sources of these complexities and to mitigate risks and difficulties associated with these systems. A high level of complexity leads to unpredictable system performance, highly variable lead time, and creates a stressful and difficult working area for operators (Calinescu et al. 2002). Manufacturing organizations must have the ability to operate systems within an ever-changing demand and environment. They must control the required operations due to different product requirements that include different process flows or routes through the facility. For example, the manufacturers needs to incorporate the ever-changing customer requests, delays in raw materials, operator skills, absenteeism, and resource unavailability (Calinescu et al. 2002). This will require an agile environment that incorporates a flexible schedule plan to handle interruptions in the system. Complexity within the manufacturing system comes from the variability of products and the uncertainty due to other factors such as material, operator, and so on. It could also be traced to the dynamic nature of the manufacturing environment that leads to an increased number of decisions to meet the future demands and the impact of these decisions (Deshmukh and Talavage 1998). The uncertainty is due to market demand, product life cycle, and so on, at the macro level and tool wear, machine breakdown, and so on, at the micro levels of operations. Figure 7.6 depicts the basic elements of the complexity.

The ability of a production line to produce a mix of product variants requires a shop floor monitoring system with integrated information flow capabilities. This system provides the required details about the locations of the product within the operation and its stages of production. Other factors may include (1) number of product types manufactured, (2) required number of manufacturing operations, and (3) required levels of system maintenance. These are managed by operator training.

Other forms of complexity are product complexity, process complexity, and operational complexity (ElMaraghy and Urbanic 2003). Product complexity is attributed to material, design, and special specifications of the components. Process complexity is based on volume requirement and the work environment that requires process decisions such as type of equipment, jigs and fixtures, tooling, and gauges. Operational complexity is based on product, process, and the production logistics. Some of the components of the operational complexity are scheduling, machine setup, monitoring and maintenance of the tasks, and so on. Figure 7.7 represents the interaction among these complexities.

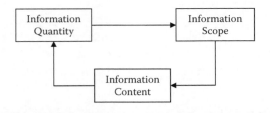

Figure 7.6 Scope of complexity.

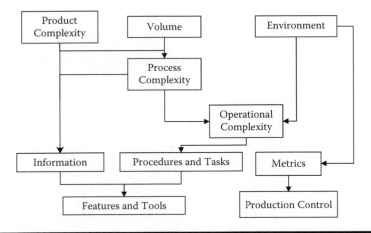

Figure 7.7 Integrated impacts of manufacturing complexity.

To better understand the overall impact of complexity on manufacturing operations, careful considerations and evaluations of the following elements are required (Calinescu et al. 1997, 2002):

■ Product structure (e.g., number of different products, product volume, lead and cycle time, lot sizes, resources required)
■ Shop or plant structure (e.g., number and types of resources, layout, setup times, maintenance, performance measures)
■ Planning and scheduling functions (e.g., scheduling strategies, decision-making process)
■ Information flow (e.g., internal, intra-plant and external suppliers, customers).
■ The variability and uncertainty of the environment (e.g., customer changes, breakdowns, data inaccuracy, unreliability)

7.5.1 Reasons of Complexity

Increased levels of manufacturing complexity could be attributed to (Wiendah and Scholtissek 1994 and Calinescu et al. 1997, 2002):

■ Market globalization: delivery time, quality, time to market, and customer satisfaction have become more important than price.
■ The coordination between order-related and customer-independent manufacturing.
■ The lack of the integration among the different segments of the system.
■ Adding flexibility to the production system without proper control.
■ Adding flexibility to the factory floor increases the scheduling alternatives and the decision-making complexity.

7.6 Product Variety and Manufacturing Complexity

The extent of product variety handled by any manufacturing firm depends on its volume of production (Kamrani et al. 2004). Low-quantity production is usually known as *job shop production*. The agility embedded in this form of production makes it the most used production system to manage high product variety, although medium-quantity production handles hard product variety using batch production. In this approach, batches of materials are pushed through different stages of production and assembly since batches are manufactured one after the other. There is a fixed time for changeover from one batch to the other. Soft product variety in medium production is handled by cellular manufacturing, where every cell specializes in the production of a given set of similar parts or products, according to the principles of group technology. Each cell is designed to produce a limited variety of part configurations. Mass production is characterized by high quantity and is dedicated to the manufacture of specific products that have high demands. Mass production can be broadly classified into quantity production and flow line production. Quantity production involves the mass production of single parts on specialized pieces of equipment. Flow line stations involve multiple workstations arranged in sequence, and the products are moved through the sequence with subassemblies assembled at respective locations. Assembly lines of automobile manufacturing are a typical example of flow line manufacturing. The term *mixed-model assembly line* refers to those assembly lines that can typically handle soft variety in the vehicle assembly operation.

Manufacturing complexity is classified into structural (static) complexity and operational (dynamic) complexity (Frizelle and Woodcock 1995). Structural or static complexity is defined as the expected amount of information necessary to describe the state of a system. Production schedule provides the data to calculate the static complexity of the manufacturing system. Static complexity is measured using the entropy equation:

$$H_s = -\sum_{i=1}^{m} \sum_{j=1}^{S} p_{ij} \log_2 p_{ij} \qquad (7.7)$$

where
  m is the number of resources.
  s is the number of scheduled states.
  p_{ij} is the probability of resource i being in scheduled state j.

Operational or dynamic complexity is defined as the expected amount of information necessary to describe the state of the system deviating from the schedule due to uncertainty. The calculation involves measurement of the difference between the actual performance of the system and the expected figures in the schedule. Dynamic complexity is given by

$$H_d = -P(\log_2 P) - (1-p)\log_2(1-p) - (1-p)\sum_{i=1}^{m} \sum_{j=1}^{ns} p_{ij} \log_2 p_{ij} \qquad (7.8)$$

where
  P is the probability of the system being in control.
  $(1-p)$ is the probability of the system being out of control.
  m is the number of resources.
  n_s is the number of nonscheduled states.
  p_{ij} is the probability of resource i being in a nonscheduled state j.

	Variety	Uncertainty
Capital	Plant & Equipment Storage IT	Excess Plant & Equipment Excess Storage Additional IT
Revenue Items	Materials Labor Utilization Transportation Overheads	Excess Labor Interest Charges Rectification Costs Revenue Loss Warranty Payments Excess Overheads

Figure 7.8   Costs generated due to complexity.

A fair estimate of the cost of increased product variety is often difficult to arrive at because variety incurs many indirect costs that are not clearly understood and are not easy to capture. Costs that are difficult to determine include raw material inventory, work in process inventory, finished goods inventory, postsales service inventory, reduction in capacity due to frequent setups, and cost of increased logistics due to added variety. Figure 7.8 represents the costs generated due to increase level of complexity (Martin and Ishii 1996).

Setup time or the batch size primarily determines the cost of variety in manufacturing. Because of large volumes, mass production has specific machinery that is relatively inflexible for handling product variants. Furthermore, mass production is often characterized by dies that have large setups, which encourage higher lot sizes. This forces mass production to have large batch sizes to minimize the downtime per product. Consequently, this results in larger work in process (WIP), larger floor space, lower quality costs, and lower machinery utilization. Work-in-process inventory costs have a direct relation with respect to the lot size in any manufacturing setup. With higher work-in-process, the production system drifts toward the push system of production. This is often accompanied by an increase in floor space utilization and increase in internal transportation costs. Large lot sizes increases the quality costs due to repeating errors, the primary reason being the increase in vulnerability to the unknowingly occurring manufacturing defects. Smaller lot sizes favor lesser part rejection and result in lower quality costs (cost of rejection). Machine utilization increases when the lot size increases because of the economy of scale. Thus, when there is a drift toward increase in product variety in mass production, machine utilization suffers. In traditional mass production, frequent setups to accommodate a wide product range also increases the setup, labor costs, and downtime. In summary, mass production is least flexible in handling product variety due to operational inefficiencies and cost increases due to product proliferation.

Author

Ali K. Kamrani is an Associate Professor of Industrial Engineering. He is also Founding Director of the Design and Free Form Fabrication Laboratory at the University of Houston, USA. He received his BS in Electrical Engineering in 1984, Master of Engineering in Electrical Engineering in 1985, Master of Engineering in Computer Science and Engineering Mathematics in 1987, and PhD in Industrial Engineering in 1991, all from the University of Louisville, Louisville, Kentucky. His research has been motivated by the fundamental application of systems engineering and its application in advanced design and development of complex systems. He is the Editor-in-Chief for the *International Journal of Collaborative Enterprise* and the *International Journal of Rapid Manufacturing.*

References

Adat, A. 2005. A Simulation based Methodology to Measure and Analyze the Required Inventory due to Product Proliferation. Master's thesis, University of Houston.

Aigbedo, H., and Y. Monden. 1996. A simulation analysis for two-level sequence scheduling for just-in-time (JIT) mixed-model assembly lines. *International Journal of Production Research* 34(11): 3107–24.

Akturk, M.S., and F. Erhun. 1999. An overview of design and operational issues of kanban systems. *International Journal of Production Research* 37(17): 3859–81.

Anderson, D.M. 1997. *Agile Product Development for Mass Customization. How to Develop and Deliver.* Chicago: Irwin Professional.

Anwar, F.M., and R. Nagi. 1997. Integrated lot-sizing and scheduling for just-in-time production of complex assemblies with finite set-ups. *International Journal of Production Research* 35(5): 1447–70.

Baykoc, O.F., and S. Erol. 1998. Simulation modeling and analysis of a JIT production system. *International Journal of Production Economics* 55(7): 203–12.

Benjaafar, S., J.S. Kim, and N. Vishwanadham. 2004. On the effect of product variety in production-inventory systems. *Annals of Operations Research* 126(1–4): 71–101.

Calinescu, A., J. Efstathiou, J. Bermejo, and J. Schirn. 1997. Assessing decision making and process complexity in manufacturing through simulation. *Proceedings of the 6th IFAC Symposium on Automated Systems Based on Human Skill.* IFAC: 159–62.

Calinescu, A., J. Efstathiou, L.H. Huatuco, and S. Sivadasan. 2002. Classes of Complexity in Manufacturing. Research Report, Manufacturing Research Group, Department of Engineering Science, University of Oxford.

Christiansen, K., J. Kjeldgaad, and K. Tvenge. 2003. Shaping Product Portfolios for the Future. Product Management Consultants, Proventia A/S.

Chu, C.H., and W.L. Shih. 1992. Simulation studies in JIT production. *International Journal of Production Research* 30(1): 2573–86.

Da Silveira, G. 1998. A framework for the management of product variety. *International Journal of Operations and Production Management* 18(3): 271–85.

Deshmukh, A., and J. Talavage. 1998. Complexity in manufacturing systems, Part 1: Analysis of static complexity. *IIE Transactions* 30: 645–55.

Detty, R.B., and J.C. Yingling. 2000. Quantifying benefits of conversion to lean manufacturing with discrete event simulation: A case study. *International Journal of Production Research* 38(2): 429–45.

Dobson, G., and C.A. Yano. 2002. Product offering, pricing, and make-to-stock/make-to-order decisions with shared capacity. *Production and Operations Management* 11(3): 293–306.

ElMaraghy, W.H., and R.J. Urbanic. 2003. Modelling of manufacturing systems complexity. *Annals of the CIRP* 52(1): 363–363.

Frizelle, G., and E. Woodcock. 1995. Measuring Complexity as an aid to developing operational strategy. *International Journal of Operations and Production Management* 15(5): 26–39.

Jina, J., A. Bhattacharya, and A. Walton. 1997. Applying lean principles for high product variety and low volume: Some issues and propositions. *Logistics Information Management* 10(1): 5–13.

Kamrani, A.K., A.K. Adat, and A. Rahman. 2004. A simulation-based methodology for manufacturing complexity analysis. *Proceedings of the National IIE Conference,* May 2004.

Lee, H., and C. Billington. 1993. Hewlett-Packard gains control of inventory and services trough design for localization. *Interfaces* 23(4): 1–11.

Li, J.W. 2003. Simulation based comparison of push and pull systems in a job-shop environment considering the context of JIT implementation. *International Journal of Production Research* 41(3): 427–47.

MacDuffie, J.P., K. Sethuraman, and M.L. Fisher. 1996. Product variety and manufacturing performance: Evidence from the international automotive assembly plant study. *Management Science* 42(3): 350–69.

Malik, S. A., and W. G. Sullivan. 1995. Impact of ABC information on Product Mix and Costing Decisions. *IEEE Transactions of Engineering Management* 42(2): 171–76.

Martin, M.V., and K. Ishii. 1997. Design for Variety: Development of Complexity Indices and Design Charts. ASME Design Engineering Technical Conferences, Sacramento, CA.

Martin, M.V., and K. Ishii. 1996. Design for variety. A Methodology for understanding the costs of product proliferation. *Proceedings of the 1996 ASME Design Engineering Technical Conferences and Computers in Engineering Conference,* August 18–22.

Martinez, F.M., and L.M.A. Bedia. 2002. Modular simulation tool for modeling JIT manufacturing. *International Journal of Production Research* 40(7): 1529–47

Mejabi, O., and S. Wasserman. 1992. Basic concepts of JIT modeling. *International Journal of Production Research* 30(1): 141–49.

Monden, Y. 1998. *Toyota Production Systems–An integrated approach to Just in Time.* London: Chapman & Hall.

Muralidhar, K., S.R. Swenseth, and R.L. Wilson. 1992. Describing processing time when simulating JIT environments. *International Journal of Production Research* 30(1): 1–11.

Prasad, B. 1997. *Concurrent Engineering Fundamentals, Vol. II: Integrated Product Development.* New Jersey: Prentice-Hall PTR.

Spedding, T.A., and G.Q. Sun. 1999. Application of discrete event simulation to the activity based costing of manufacturing systems. *International Journal of Production Economics* 58: 289–301.

Tang, E., and R. Yam. 1996. Product variety strategy—an environmental perspective. *Integrated Manufacturing Systems* 7(6): 24–29.

Wiendah, H., and P. Scholtissek. 1994. Management and control on complexity in manufacturing. *Annals of the CIRP* 43(2): 533–40.

Yeb, K., and C. Chu. 1991. Adaptive strategies for coping with product variety decisions. *International Journal of Operations and Production Management* 11(8): 35–47.

Chapter 8

A Simulation-Based Methodology for Manufacturing Complexity Analysis

Ali K. Kamrani,[1] Arun Adat,[2] and Maryam Azimi[3]

[1]*Industrial Engineering Department, University of Houston, Houston, Texas*

[2]*Hewlett Packard, Houston, Texas*

[3]*Lenovo Corporation, Morrisville, North Carolina*

Contents

Abstract

This chapter presents a sample case study for analyzing manufacturing complexity due to increased product variety. A cost model is developed that captures the impact of increased inventory and the storage cost of subassemblies due to increased product variants. This is accomplished by generating a mixed-model assembly sequence that aims to minimize the variation of subassembly inventories of the production span. The mixed-model assembly line is simulated to trace the inventory levels of the individual subassemblies. The output of the simulation model is used in developing the cost models to give the daily cost of inventory holding and storage of the different subassemblies.

8.1 Mixed-Model Assembly Analysis

The inventory level in a mixed-model assembly line is primarily determined by the sequence of the components that are assembled on the line. If products with longer processing times are successively fed into the assembly lines, a delay in model completion will eventually occur (Monden 1998). To prevent this from occurring, the processing time at each station must be managed by sequencing models so that, in general, a model with a relatively short processing time at a station follows soon after a model with a relatively long processing time. Furthermore, the quantity of each part used per unit time must be as near constant as possible because it is crucial for the processes preceding the assembly line supplying components to have a uniform demand (Akturk and Erhun 1999; Anwar and Nagi 1997; Chu and Shih 1992). The uniform demand allows the JIT (just-in-time) "pull" system to minimize work-in-process inventories. Researchers compared the goal chasing method and the goal chasing method II developed at Toyota, Miltenburg's algorithm, and time spread methods of sequencing for the assembly line, efficiency factors such as work not completed, worker idleness workerstation time, and a measure of variability in uniform component usage (Miltenburg 1989). Their study showed that time spread and Miltenburg's algorithms were the most effective overall sequencing procedures. Time spread was preferable if assembly line efficiency was considered, whereas Miltenburg's algorithm performed better if part usage was given primary importance. Aigbedo and Monden experimentally investigate the effect on subassembly usage smoothening when the product usage smoothening is considered together with the subassembly usage smoothening goal for determining the assembly line product sequence (Aigbedo and Monden 1996). The impetus for their study was Miltenburg's remark that Toyota neglected product usage smoothening and only considered part usage smoothening while determining the assembly line product sequence. The results of their study showed that two-level scheduling was computationally faster but in general performed poorer than single-level smoothening. Drexl and Kimms formulated the JIT mixed-model assembly line sequencing problem as an integer program considering both part usage constraints and station load constraints (Drexl and Kimms 2001). Their results showed that solving the LP relaxation of the problem by column generation provides tight lower bounds for the optimal objective function value. Garcia and Sabater claimed that mixed-model assembly sequencing considering leveled component consumption, option appearance, and smoothened workload is dynamic in nature because of the availability of the different products to be sequenced (Garcia-Sabater 2001). They also propose and test an efficient parametric procedure that adapts to the varying conditions. Mane et al. tested Toyota's Goal Chasing Algorithm I and user-defined algorithm on an Australian automotive company (Mane et al. 2002). They concluded that the goal chasing algorithm generated sequences better than the user-defined algorithm although user-defined algorithm was flexible in accommodating user-defined priorities. Simulation has been conventionally used to model assembly lines due to the stochastic nature of assembly operations. Muralidhar et al. described process times in a JIT environment using truncated normal, gamma, and lognormal distributions and the concluded that gamma distribution was the best suited for describing processing times (Muralidhar et al. 1992). Mejabi and Wasserman suggested that during JIT implementation some subsystems will continue to retain their push characteristics (Mejabi and Wasserman 1992). They proposed a control paradigm based on the concept of kanban satisfaction that provides a control structure that permits the pulling of material to take place in a JIT environment. They also claimed that, of the high-level languages, only WITNESS* possessed any real JIT capabilities. Carlson and Yao compared the push and the pull production systems in handling mixed-model assembly operation (Carlson and Yao 1992). Their results showed that the pull production system performed better than the push production system if the assembly line rejects were low. Higher reject rates in JIT assembly lowered the line performance greatly due to the absence of queues between the stations. They also showed that push systems perform significantly better than pull systems if the defect rates are high. Chu and Shih emphasized the use of simulation in analyzing JIT production systems (Chu and Shih 1992). They proposed simulation as a successful tool in evaluating factors such as a measure of the company's JIT performance, acceptable inventory level, and use of two cards or single card kanban. They claimed that many simulation-related statistics were ignored in previous studies and conclusions drawn from previous studies need to be reconfirmed. They also felt that most researchers had a common perception that some experimental factors were more important than the others, and the overall behavior of these factors has not yet being well explained. Wang and Xu tested the performance of a hybrid push/pull production using strategy simulation software for flow shop manufacturing (Wang and Xu 1997). The model simulated the material flow for different production strategies and the simulation results demonstrated that the recommended push/pull strategy was the best for the general mass product manufacturing systems. Baykoc and Erol model a multi-item, multistage JIT system in SLAM II, a FORTRAN-based simulation language, and analyze the effects of increase in the number of kanbans on production performance (Baykoc and Erol 1998). Their results showed that output rate and utilization are increased as the number of kanbans increase, but no improvement is observed after two kanbans. Also, increasing the number of kanbans result in a striking increase in waiting times and work in process lengths. Hence, they concluded that the ideal number of kanbans was two for the system considered in their study. Their results also showed that better performance on a mixed-model JIT system depends on reducing or eliminating (if possible) variations related to assembly time, demand arrivals, and balance between stations. Savsar simulates assembly of printed circuit board and analyzes push and pull production systems (Savsar 1997). His study shows that the simulation modeling approach can be utilized to determine the minimum kanbans needed to circulate in the system and the WIP buffer levels needed to meet a specified percentage of demand on time in a real assembly line setting. Bukchin studied the throughput of a mixed-model assembly line using six performance measures through simulation (Bukchin 1998). The performance measures include smoothed station measure, minimum idle time measure, station's coefficient of variation, bottleneck measure, and model variability measure. The bottleneck measure was the best measure in almost all simulation results followed by model variability and smoothed station. Spedding and Sun used discrete event simulation to evaluate the activity-based costs of a manufacturing system (Spedding and Sun 1999). Witness* simulation software was used to model a semiautomated printed circuit board assembly line. Under similar conditions, the simulation model gave the same estimates as those derived from the IDEF modeling approach. However, simulation models had the advantage of being able to provide greater detail and take into account the intrinsic variation of a dynamic manufacturing system. Akturk and Erhun classified techniques to determine both the design parameters and kanban sequences for just-in-time manufacturing (Akturk and Erhun 1999). They observed that JIT is based on repetitive manufacturing. Therefore, factors that adversely affect the repetitive nature of the system such as the increasing the product variety and decreasing product standardization reduce the performance of kanban systems. It was also observed that perfectly balanced lines outperform the imbalanced ones even when we vary the number of kanbans at each stage. Detty and Yingling used discrete event simulation to quantify the benefits of conversion to lean manufacturing (Detty and Yingling 2000). Their results showed an average reduction in waiting time of parts, model change-over times, reduction in floor space, and average inventory levels in the lean system as compared to the existing system. A noteworthy result was that the lean system had an 86% reduction in average lead time compared to the traditional manufacturing system. Martinez and Bedia use the modular capabilities of Witness* to introduce a modular simulation tool (Martinez and Bedia 2002). They build a U-shaped line by integrating a feeder double kanban line module. Their studies demonstrate the use of modular capabilities of Witness˚ in analyzing system configurations and scheduling rules before implementing them. Li used a simulation-based approach to compare push and pull production environments considering the context of JIT implementation (Li 2003). They found that setup reduction effected by cellular manufacturing substantially affected the one-piece production and conveyance in job shop environment. Tables 8.1 and 8.2 summarize the results of some methods used.

Table 8.1 Summary of Studies in Mixed-Model Assembly Sequencing

Researchers	Work	Findings
Monden (1983)	Goal Chasing Heuristic	Advantages and disadvantages of Goal Chasing heuristics
Sumichrast, Russell, and Taylor (1992)	Compare Goal chasing heuristic I and II, time sharing and Miltenburg's algorithm	Miltenburg's algorithm was the most effective overall heuristic followed by time sharing heuristic
Monden and Aigbedo (1996)	Compare subassembly level smoothening and combination of subassembly level with product-level smoothening	Two-level smoothening was computationally faster but in general performed poorer than single-level smoothening
Rodriguez, Garcia, and Lario (1999)	New objective function to solve mixed-model sequencing by genetic algorithm	Genetic algorithm was superior in the quality of sequence but had higher computational solving time
Garcia and Sabater (2001)	Dynamic nature of the sequencing problem when considering leveled component consumption, option appearance, and smoothened workload	Develop an efficient parametric heuristic that adapts to these varying conditions
Drexl and Kimms (2001)	Mixed-model sequencing as an integer program	LP relaxation by column generation provides tight lower bounds for the objective function

Table 8.2   Simulation in Modeling Mixed-Model Assembly Lines

Researchers	Work	Findings
Muralidhar, Swenseth, and Wilson (1992)	Process times in JIT environment	Gamma distribution was best suited for describing process times
Mejabi and Wasserman (1992)	Control paradigm based on kanban satisfaction	Witness® possessed real-time JIT capabilities
Carlson and Yao (1992)	Comparison of push and pull production systems in mixed-model assembly line	Pull performed better than the push production system if assembly line rejects were low
Wang and Xu (1992)	Hybrid push/pull production systems	Hybrid push/pull production system was the best for mass product manufacturing systems
Baykoc and Erol (1998)	Effects of increase in the number of kanbans on production performance	Ideal numbers of kanbans were two for their system of study
Bukchin (1998)	Throughput of a mixed-model assembly line using six performance measures	Bottleneck measure was the best measure followed by model variability and smoothed station
Spedding and Sun (1999)	Discrete event simulation for activity-based costing of manufacturing systems	Simulation model gave same results to IDEF model along with added flexibility
Martinez and Bedia (2002)	Modular capabilities of Witness® to introduce a modular simulation tool	Built a U-shaped line by integrating a feeder double kanban line module
Li (2003)	Simulation-based methodology to compare push and pull in the context of JIT implementation	Setup reduction effected by cellular manufacturing substantially effected one-piece production and conveyance in job shop environment

8.2 Impact of Product Variety on Manufacturing Costs

The ideal product variety to offer and the cost of added product variety has been approached differently by researchers in the past. Malik and Sullivan used mixed integer programming, which utilizes activity-based cost (ABC) information to determine the optimal product mix and product cost in a multiproduct manufacturing environment (Malik and Sullivan 1995). They showed with an example that with the traditional costing approach, it was possible to arrive at a product mix that may not be achievable with a given capacity of indirect resources. Furthermore, adopting a product mix strategy suggested by traditional costing methods might also increase the overhead costs, which are not anticipated in the early stages of planning and costing. MacDuffie et al. examined the effect of product variety on manufacturing performance (MacDuffie et al. 1996). The performance factors include total labor productivity and consumer-perceived product quality. They define three dimensions of product variety: fundamental, peripheral, and intermediate variety. Their study supports the hypothesis that lean production plants are capable of handling higher levels of product variety with less adverse effect on total labor productivity than traditional mass production plants. Their study partly explains how the leanest Japanese plants have been able to achieve higher overall performance with much higher levels of parts complexity and option variability. Ishii and Martin introduce the concept of design for variety (DFV), which is a tool that enables product managers to estimate the cost of introducing variety into their product line (Ishii and Martin 1996). They claimed that cost estimates used to determine the profitability of the companies that offered new product offerings did not account for all the costs associated with providing this additional variety. Their model attempts to capture the indirect cost of variety through the measurement of three indices: commonality, differentiation point, and setup cost. DFV methodology was a basic procedure for helping managers and engineers understand the true costs of introducing variety into their product line. Benjaafar et al. examined the impact of product variety on inventory costs in a production inventory system with finite capacity assuming make-to-stock production, setup times, finite production rate, and stochastic production times (Benjaafar et al. 2004). Their results show that inventory costs increase linearly with the number of products. They also show that the rate of increase is sensitive to system parameters, including demand and process variability, demand and capacity levels, and setup times. Dobson and Yano formulate the problem of product variety and pricing as a nonlinear program (Dobson and Yano 2002). They assume a manufacturer who has a single machine or production line that is capable of producing a range of potential products. The effect of inventory costs associated with the products is captured by modeling the time between production runs as a decision variable. Their results show that the optimal product mix depends strongly on the production cycle duration. Ozbayrak et al. estimated the manufacturing costs of an advanced manufacturing system that runs under a material requirement planning (MRP) or just-in-time (JIT) system by using activity-based costing (Ozbayrak et al. 2004). They use a simulation modeling tool to observe the manufacturing cost behavior under two separate control strategies, the push and pull systems. Parts are either pushed or pulled and are sequenced according to the four scheduling rules: shortest processing time, longest processing time, first come first serve, and slack. They found that randomness, buffer capacity, and lead times are found to be important cost drivers in terms of their effect on work in progress (WIP) and throughput, and an increase in variation and buffer capacity can result in a buildup of WIP inventory and a slight increase of throughput volumes with the expense of considerable increase in manufacturing costs. Table 8.3 summarizes the results of past works on product variety and their impacts on manufacturing complexity.

8.3 Case Study Overview

The research site for the project is the axle assembly operation of a major automobile company that assembles a variety of vehicle models. Axles and coils are delivered daily based on the scheduled production, and a level of safety stock is always maintained in the event of any out-of-equence production (e.g., missing components or new production schedule). Rear and front axles are installed onto the vehicle chassis on a moving platform at the first two stations and then moved through the line for other tasks such as brake line and spring coil assembly. The following are the different subassembly variants of front/rear axles and front/rear coil springs in the assembly line:

■ 9 front axles: 184AP, 184AQ, 187AQ, 187AR 600AC, 600AD, 601AC, 601A, and 601AE
■ 4 rear axles: 426AG, 429AG, 430AF, and 433AG
■ 7 rear spring coils: 344, 345, 400, 404, 500, 550, and 551
■ 7 front spring coils: 262, 263, 264, 265, 267, 268, and 269

Based on the above list, there are 1,764 product combinations of front/rear axles and front/rear spring coils possible, although only 55 vehicle models are built due to design and operational qualifiers. For instance, a heavy-duty axle cannot be matched with a low-stiffness spring. A PSG (product structure graph) representation of the possible axle and coil combinations is shown in Figure 8.1. The assembly line has an operation cost (inventory holding and storage) corresponding to the current inventory level. The objective is to determine the variation in inventory level and the corresponding operational costs if an additional model (product variant) is introduced into the manufacturing operations. This study focuses on the possible vehicle variants based on various axles and spring coils combinations. An overview of the general implementation approach and methodology is illustrated in Figure 8.2.

Table 8.3 Impact of Product Variety on Manufacturing

Researchers	Work	Findings
Malik and Sullivan (1995)	Mixed integer programming that utilizes activity-based cost information	Traditional costing approach gives infeasible product mixes and gives unanticipated overhead costs in planning and costs
MacDuffie, Sethuraman, and Fisher (1996)	Effect of product variety on manufacturing systems	Lean production plants are capable of handling higher product variety with less adverse effects on total labor productivity
Ishii and Martin (1996)	Introduce design for variety, which is a tool that enables product managers estimate the cost of product variety	Earlier cost estimates did not account for all the costs associated with providing additional variety. DFV helps managers understand the true costs of introducing product variety
Benjafar, Kim, and Vishwanadham (2002)	Impact of product variety on inventory costs in a production system working on make-to-stock policy	Inventory costs increase linearly with the number of products and the rate of increase is sensitive to the system parameters such as demand and process variability, demand and capacity levels, and setup times
Dobson and Yano (2001)	Formulate the problem of variety as a nonlinear program	Optimal product mix depends strongly on the production cycle duration
Ozbayrak, Akgun, and Turker (2004)	Estimate the cost of manufacturing of an advanced manufacturing system, running on MRP or JIT system by using ABC methodology	Randomness, buffer capacity, and lead times are the important cost drivers on WIP and throughput
Kamrani, Adat, and Rahman (2004)	Model to analyze and mitigating the risks associated with product variety and its impact on manufacturing complexity	In-house sequencing performs better than push production system

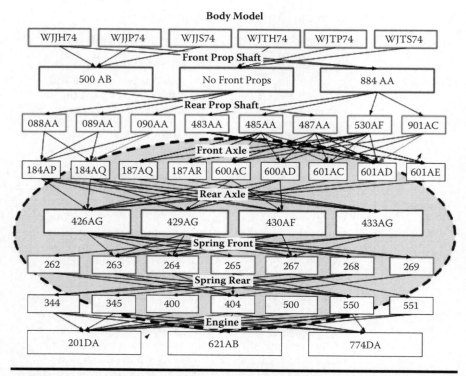

Figure 8.1 Schematic view of the possible combinations of subassemblies.

8.3.1 Simulation-Based Methodology

Prior to the development of the cost impacts, part delivery and storage strategies must be determined. Three possible scenarios are considered. These are push delivery, sequencing delivery, and in-house sequencing delivery models. Currently, axles are delivered daily based on the production schedule, and a level of safety stock is always available. This safety stock level is estimated by the axle supplier. Out-of-sequence events are due to wrong part sequence, added schedule, and other situations such as breakdown, part damage, and so on. Wrong axles are transferred manually by the line operator and placed in the pull-off bins.

In the case of out-of-sequence scenarios, safety stock bins are searched until the right axle is located and then assembled on the platform. Excess inventory is stored at various locations in the plant. The objective of this is to minimize the area required used for storing of the excess inventories and to utilize this area for other value-added operations.

Rear and front axles are installed onto the moving platform at the first two stations and then moved through the line for other tasks such as break line and coil assembly. Figure 8.3 illustrates the area where axles and the coils are assembled into

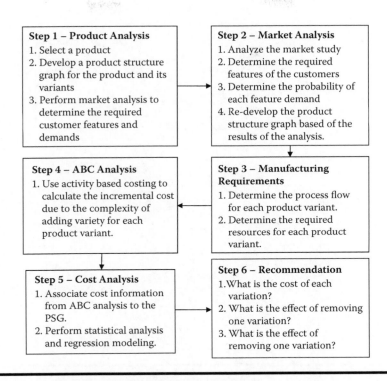

Step 1 – Product Analysis
1. Select a product
2. Develop a product structure graph for the product and its variants
3. Perform market analysis to determine the required customer features and demands

Step 2 – Market Analysis
1. Analyze the market study
2. Determine the required features of the customers
3. Determine the probability of each feature demand
4. Re-develop the product structure graph based of the results of the analysis.

Step 4 – ABC Analysis
1. Use activity based costing to calculate the incremental cost due to the complexity of adding variety for each product variant.

Step 3 – Manufacturing Requirements
1. Determine the process flow for each product variant.
2. Determine the required resources for each product variant.

Step 5 – Cost Analysis
1. Associate cost information from ABC analysis to the PSG.
2. Perform statistical analysis and regression modeling.

Step 6 – Recommendation
1. What is the cost of each variation?
2. What is the effect of removing one variation?
3. What is the effect of removing one variation?

Figure 8.2   Scope of the proposed methodology.

the platform. Figure 8.4 illustrates the methodology used for implementing this stage of the analysis. The implementation steps are

- Data collection and verification
- Simulation model development and verification
- Base model complexity analysis (part mix complexity analysis—structural and manufacturing mix analysis—dynamic)
- Delivery/inventory policy

Table 8.4 lists the required data for the development of simulation models. Three different simulation models are developed.

The first model is based on the current operation of the plant using the sequenced deliverer policy. For this problem, only part-mix complexity (structural) and dynamic complexity (WIP) are measured. The part-mix complexity measure is the same ($7.3 \approx 8$ bits) for all three models, although dynamic complexity significantly varies from one model to another. Figure 8.5 illustrates the measure of dynamic complexity for the sequenced mode of plant operations. From this analysis, the dynamic measure begins to stabilize at 60%. This is an indicator that

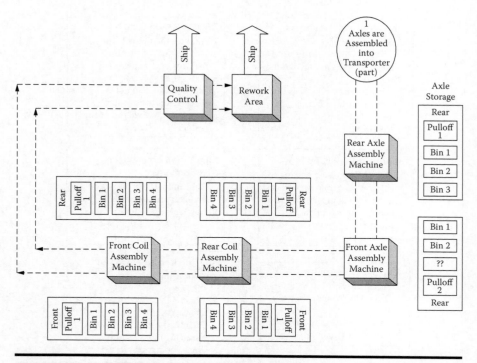

Figure 8.3   Operational sequence and assembly flow.

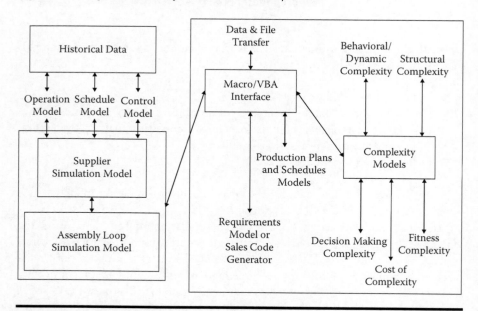

Figure 8.4   Scope of the proposed system.

Table 8.4   Required Data for the Simulation Model

Required Information	Mode Type
Production mix data	Front and rear RS Spring
	Front and rear LS Spring
	T-Case and engine type
	Rear and front prop
	Rear and front axle
Pull off data	Number of pull off
	Type of axle
	Frequency
	Reason
Delivery information	Delivery schedule
	Number of delivery per day for the axles
	Type of axles
	Number of hi–low travel moving/removing per day
	Number of hi–low
	Number of delivery per day for springs
	Type of springs
	Number of spring per batch
	Number of hi–low travel moving/removing per day
Cycle and stoppages data	Cycle time at each station thin axle loop
	Number of downtime at each station
	Mean time to failure (MTTF)
	Mean time to repair (MTR)
	Reason for failure
	Number of setup
	Setup time at each station

(Continued)

Table 8.4 (*Continued*)   Required Data for the Simulation Model

Required Information	Mode Type
Repair data	Number of repairs due to wrong axles install (done at axle loop)
	Number of repairs due to wrong axles install (not done at the axle loop)
	Repair time for each case
	Delay time before maintenance schedule
	Other repairs due to the process done at the axle loop
	Repair time
	Number of repairs due to wrong spring install (done at the axle loop)
	Number of repairs due to wrong spring install (not done at the axle loop)
	Repair time for each case
	Delay time before maintenance schedule
Facility data	Square footage for the WP inventory for the springs
	Square footage for the WP inventory for the axles
	Sq/ft cost

the system is going through many dynamic changes. This is determined to be due to the changes in production schedules that require axles to be pulled off during assembly. The dynamic measures for push and in-house sequence are lower since they are mainly due to production stoppages (breakdown, repair, etc.). In these two cases, no parts are out of sequence. The only impact associated with these two policies is the required inventory. Figure 8.6 illustrates the cost impacts and comparisons among all three policies due to complexity. The cost is calculated based on

- Cost of axle delivery
- Cost of axle storage
- Cost of fork-lift operator
- Cost of pull-off inventory
- Cost of resequencing
- Cost of operators
- Other costs (overtime, etc.)

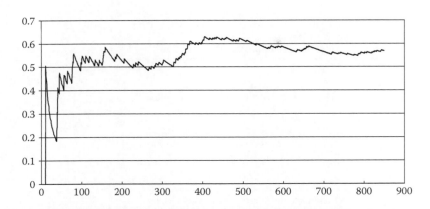

Figure 8.5 Dynamic measure of complexity for the current scenario for the total built.

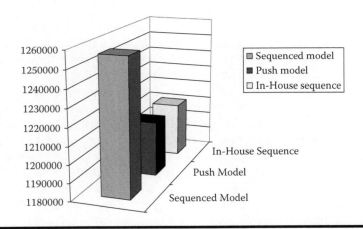

Figure 8.6 Cost comparisons between different delivery and inventory policies.

Based on the simulation results, the current system setup is well designed to handle the current combinations of the vehicle assembly, although the impact of complexity is significant on material handling and inventory costs.

From Figure 8.6, parts delivered and then sequenced in-house would result in a significant lower cost and saving to the company. The next step of the analysis is to develop a comprehensive cost model used to study the impact of added product variety. This is captured by a sequence-driven simulation model as shown in Figure 8.7.

A three-phase solution methodology is proposed to develop the cost model to quantify the effects of added product variety. In the *Sequence Generation* phase,

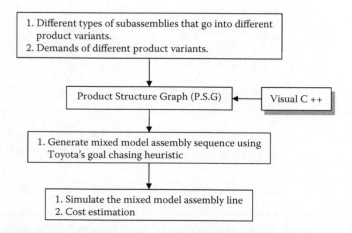

Figure 8.7   The proposed cost model for added variety analysis.

a mixed-model assembly sequence is generated using Toyota's goal-chasing algorithm. The goal-chasing algorithm is coded in Visual C++, and it only considers single-level subassembly smoothing. It is assumed that the monthly demands of the vehicle models are known. By knowing subassembly requirements corresponding to the vehicle models, the goal-chasing algorithm generates the mixed-model assembly sequence aiming to minimize the variation in consumption of the subassemblies. In the *Assembly Line Simulation* phase, the assembly line is simulated using Witness® simulation software.

The mixed-model assembly sequence generated by the goal-chasing algorithm is used as input to the simulation model. Subassemblies are replenished at the beginning of every shift. It is assumed that worker productivity of the assembly line is unaffected by the assembly sequence, and there is no shortage of the vehicle models that are scheduled by the goal-chasing algorithm. The assembly sequence follows the JIT methodology, and there is no buffer between the assembly stations. The productivity of the line is primarily governed by the availability of the subassemblies that fit on the vehicle models. The average inventory levels of the subassemblies and the assembly utilization rates are monitored. These parameters from the simulation model are used in the third phase to compute the inventory holding and storage cost of increased product variety in the manufacturing system. Inventory holding cost is a function of the average inventory level and the annual cost of inventory holding. The storage cost is governed by the total number of bins required to store the axles and the spring coils. The numbers of bins determine the required storage space, and the company incurs a storage cost of $40.00/sq. ft./day. The central premise of the cost model is that added product variants cause a variation in the mixed-model sequence and, consequently, the inventory level of the subassemblies. This will be successfully captured by the simulation model and will be reflected in the cost model.

8.3.2 Construction of PSG for the Manufacturing System

Given the monthly production schedule, the daily production is calculated assuming that there are 22 production days in a month. From the daily production requirement, the consumption rates of the subassemblies are calculated assuming that production comprises two shifts of 435 minutes each. The PSG is shown in Table 8.5.

8.3.3 The Goal Chasing Method (GCM)

The assembly sequence in a mixed-model assembly line could be generated by the goal-chasing heuristic (Monden 1998). The primary goal of the heuristic is to minimize the variation of consumption of subassemblies that feed the main assembly line. The following are the notations pertaining to the algorithm:

- Q = Total production quantity of all products $A_j(i = 1,2,\ldots,\alpha)$
- N_j = Total necessary quantity of the part a_j to be consumed for producing all products $A_i(i = 1,2,\ldots,\alpha; j = 1,2,\ldots,\beta)$
- X_{jk} = Total necessary quantity of the part a_j to be utilized for producing the products of determined sequence from first to k^{th}.
- b_{ij} = Necessary quantity of the part $a_i(j = 1,2\ldots,\beta)$ for producing one unit of the product $A_i(1,2,3\ldots,\alpha)$

The steps are as follows:

- *Step* 1: Set K = 1, $X_{j,k-1} = 0$, $(j = 1,2,\ldots,\beta)$, $S_{k-1} = (1,2\ldots,\alpha)$
- *Step* 2: Set as Kth order in the sequence schedule the product A_j that would minimize the distance D_k. The minimum distance will be found by the Equation 8.1: $D_{ki}^{*} = min\{D_{ki}\} i \in S_{k-1}$, where

$$D_{ki} = \sqrt{\sum_{j=1}^{B}(kN_j / Q - X_{j(k-1)} - b_{ij})^2} \qquad (8.1)$$

- *Step* 3: If all units of a product A_i were ordered and included in the sequence schedule, then Set $S_k = S_{k-1} - \{i^*\}$. If some units of a product A_j are still remaining as being not ordered, then set $S_k = S_{k-1}$
- *Step* 4: If $S_k = \varphi$ (empty set), the algorithm will end. If $S_k = 0$, then compute $X_{jk} = X_{j,k-1} + b_{i^*j}(j = 1,2,\ldots,\beta)$ and go back to Step 2 by setting $K = K + 1$. As an example of the application of GCM, consider three vehicle models A_1, A_2, and A_3 required in quantities 2, 3, and 5, respectively, assuming all three models require subassemblies a_1, a_2, a_3, and a_4. The requirements of models and subassemblies are given in Table 8.6.

Table 8.5 PSG of the Manufacturing System for the Monthly Production Schedule

M	Demands	Front Axle									Rear Axle				Front Spring							Rear Spring						
		184AP	184AQ	187AQ	187AR	600AC	600AC	601AC	601AD	601AE	426AC	429AC	430AF	433AC	262	263	264	265	267	268	269	344	345	400	404	500	550	551
M1	5561				1									1		1									1			
M2	2434			1										1		1									1			
M3	2433		1											1	1												1	
M4	1651																								1			
M5	1604						1		1		1						1								1			
M6	1484					1						1					1								1			
M7	1424					1							1			1									1			
M8	1367									1		1			1		1								1			
M9	841	1									1			1														
M10	767						1						1			1			1						1		1	
M11	701				1									1														1
M12	397									1	1									1			1					
M13	369								1		1									1			1					

(Continued)

343	M14		1					1					1		1					
325	M15	1								1		1		1						1
322	M16				1		1				1							1		
320	M17				1		1					1					1			
313	M18	1		1						1		1	1						1	1
312	M19		1											1	1					
267	M20			1				1				1					1			
250	M21				1	1		1					1				1			
235	M22			1									1							1
233	M23			1				1			1							1		
223	M24			1				1	1					1			1			
201	M25				1		1						1				1			
187	M26				1	1								1		1				
181	M27			1						1		1								
170	M28			1						1		1							1	

Table 8.5 (Continued) PSG of the Manufacturing System for the Monthly Production Schedule

Demands		Front Axle									Rear Axle				Front Spring							Rear Spring						
		184AP	184AQ	187AQ	187AR	600AC	600AC	601AC	601AD	601AE	426AG	429AG	430AF	433AG	262	263	264	265	267	268	269	344	345	400	404	500	550	551
130	M29							1			1						1								1			
123	M30									1	1									1		1						
114	M31	1	1										1		1										1			
111	M32			1							1						1								1			
101	M33																1								1			
101	M34						1					1		1					1			1						
98	M35						1					1							1	1		1						
91	M36			1										1														1
80	M37	1									1	1					1								1		1	
78	M38		1														1											
76	M39			1										1					1			1						

M40	M41	M42	M43	M44	M45	M46	M47	M48	M49	M50	M51	M52	M53	M83/54	M55
1			1							1					
	1					1							1		
		1		1			1	1	1		1			1	
															1
											1				
					1										
													1		
											1				1
1			1		1					1					
						1									
	1	1													
				1		1		1	1		1		1		
	1			1		1							1		
1		1	1			1			1						
							1	1		1		1			
											1			1	1
															1
1										1					
		1	1		1						1	1			
			1		1		1	1							
	1					1						1	1		
76	69	69	67	64	62	59	55	52	45	45	42	39	38	1	34

Table 8.6   Production Quantities and Parts Conditions b_{ij}.

Products A_i	Parts a_j			
	a_1	a_2	a_3	a_4
A_1	1	0	1	1
A_2	1	1	0	1
A_3	0	1	1	0

The total necessary quantity (N_j) of part a_j ($j = 1,2,3,4$) for producing all products A_i ($i = 1,2,3$) can be computed as follows (8.2):

$$[N_j] = [Q][b_{ij}] = [2,3,5] * \begin{bmatrix} 1011 \\ 1101 \\ 1110 \end{bmatrix} \tag{8.2}$$

Further, the total production quantity of all products A_i ($i = 1,2,3$) will be

$$\sum_{i=1}^{3} Q_i = 2 + 3 + 5 = 10 \tag{8.3}$$

Therefore, $[N_j/Q] = [5/10, 8/10, 7/10, 5/10]$ and ($j = 1,2,3,4$).

Applying the values of $[N_j/Q]$ and $[b_{ij}]$ to the formula in step 2 of the above algorithm, when $K = 1$, the distances D_{ki} can be computed as follows:
For $i = 1$,

$$D_{1,1} = \sqrt{((1.5/10) - 0 - 1)^2 + ((1.8/10) - 0 - 0)^2 + ((1.7/10) - 0 - 1)^2 + ((1.5/10) - 0 - 1)^2}$$

or = 1.11
For $i = 2$,

$$D_{1,2} = \sqrt{((1.5/10) - 0 - 1)^2 + ((1.8/10) - 0 - 1)^2 + ((1.7/10) - 0 - 0)^2 + ((1.5/10) - 0 - 1)^2}$$

or = 1.01
For $i = 3$,

$$D_{1,3} = \sqrt{((1.5/10) - 0 - 0)^2 + ((1.8/10) - 0 - 1)^2 + ((1.7/10) - 0 - 1)^2 + ((1.5/10) - 0 - 0)^2}$$

or = 0.79
Thus, $D_1, i^* = \min\{1.11, 1.01, 0.79\} = 0.79$ $\therefore i^* = 3$

Therefore, the first order in the sequence schedule is the product A_3. Next, by proceeding to step 4 of the algorithm:

$X_{jk} = X_{j,k-1} + b_{3j}$
$X_{1,1} = 0 + 0 = 0$
$X_{2,1} = 0 + 1 = 1$
$X_{3,1} = 0 + 1 = 1$
$X_{4,1} = 0 + 0 = 0$
Next, when $k = 2$, then

For $i = 1$,

$$D_{2,1} = \sqrt{((2.5/10)-0-1)^2 + ((2.8/10)-1-0)^2 + ((2.7/10)-1-1)^2 + ((2.5/10)-0-1)^2}$$

or $= 0.85$
For $i = 2$,

$$D_{i,2} = \sqrt{((2.5/10)-0-1)^2 + ((2.8/10)-1-1)^2 + ((2.7/10)-1-0)^2 + ((2.5/10)-0-1)^2}$$

or $= 0.57$
For $i = 3$,

$$D_{i,3} = \sqrt{((2.5/10)-0-0)^2 + ((2.8/10)-1-1)^2 + ((2.7/10)-1-1)^2 + ((2.5/10)-0-0)^2}$$

or $= 1.59$

Thus, $D_{2,i^*} = $ Min $\{0.85, 0.57, 1.59\} = 0.57 \therefore i^* = 2$
Thus, the second order in the sequence is product $A2$. Also X_{jk} will be calculated as follows:

$X_{jk} = X_{j,k-1} + b_{2j}$
$X_{1,2} = 0 + 1 = 1$
$X_{2,2} = 1 + 1 = 2$
$X_{3,2} = 1 + 0 = 1$
$X_{4,2} = 0 + 1 = 0$

As a result, the complete sequence schedule of this example will be A_3, A_2, A_1, A_3, A_2, A_3, A_3, A_1, A_2, A_3.

8.3.4 Simulation Modeling—Mixed-Model Assembly Sequence

For the daily production sequence shift replenishment model, the mixed-model assembly sequence is used as input into the simulation model. The vehicle models are pulled in based on the input sequence rules. The outputs are sent to the

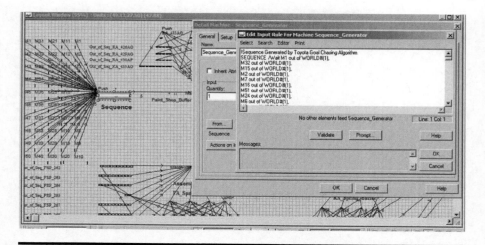

Figure 8.8 Mixed-model assembly input sequence to the simulation model.

shop buffer. This is the static buffer, which is the starting point of the axle, spring coil assembly system. Since the monthly demand is targeted at 26,835 vehicles, theoretically there are 26,835 positions in the sequence. Hence, a part file is created with the assembly sequence and linked to the simulation input file processor. An example of the simulation model is illustrated in Figure 8.8.

8.3.5 Cost Estimation

The inventory costs associated with the increased product variety comprises the cost of inventory holding and storage. The inventory holding and the storage costs are calculated by monitoring the average daily inventory level of all the subassemblies. The daily total cost of operation is estimated by the Equation 8.4:

$$C_{day} = \sum_{i=1}^{n} C_i h A_i + \sum_{i=1}^{n} (SB_i) \qquad (8.4)$$

where
   C_i = Cost of the axle (i)
   h = Inventory holding cost as a percentage
   A_i = Average inventory level of subassembly (i) per day
   B_i = Number of storage bins required for subassembly (i)
   S = Storage cost per square feet per day
   n = Total number of subassemblies in the system.

The inventory holding cost is calculated as the product of the average inventory holding and the daily average inventory level per subassembly. The axles and spring coils are assumed to be stored in storage bins. Each bin can store 10 axles (front/rear) or 192 spring coils (front/rear). One bin occupies a storage space of 8.89 ft². It is assumed that subassemblies and safety stock are stored in separate bins. There are no common bins that store multiple subassemblies.

8.3.6 Experimentations and Results

The cost impact of increased product variety was studied in two production models:

- Daily production sequence shift replenishment model
- Monthly production sequence hourly replenishment model

In the daily production sequence model, the production sequence is generated by averaging the monthly demand of the vehicle models and estimating daily demands. The daily sequence generated is repeated every day over the span of 22 consecutive days. The replenishment of all the subassemblies is done at the beginning of every shift. In the monthly production sequence model, the monthly demands of the vehicle models are used to determine the production sequence. The resulting sequence is continuous, and the replenishment of most of the subassemblies is done on an hourly basis. Those subassemblies that have a low consumption rate are replenished at the beginning of every shift to reduce the work-in-process inventory, handling costs, and line stoppages.

8.3.6.1 Daily Production Sequence with Shift Replenishment

The average inventory levels for subassemblies plotted for four different scenarios are illustrated in Figures 8.9 through 8.12. The base scenario (0,0) has 55 product variants with a total production of 1,249 vehicles per day. The second scenario (0,1) has a new product variant (56th vehicle), which uses rear axle 429 AG, front axle 184 AQ, rear spring coil 500, and front spring coil 263. The new production requirement is 1,250 vehicles per day. The third scenario (1,0) introduces the 57th product variant but removes the 56th product variant. The 57th variant uses rear axle 429 AG, front axle 184 AP, rear axle 500, and front 265. The production rate still remains at 1,250 per day. The fourth scenario (1,1) includes both the vehicle variants and, hence, the production rate is 1,251 vehicles per day.

The storage cost per day for all the subassemblies is calculated and summarized in Figure 8.13. The graph shows that there is a steady increase in the storage cost of front and rear axles, but there is only a marginal increase in the storage cost of rear spring coils and no increase in the storage cost of front springs.

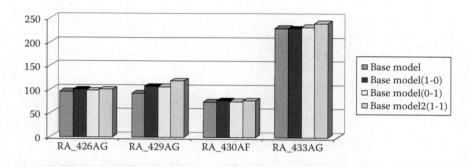

Figure 8.9   **Average daily inventory level of rear axles.**

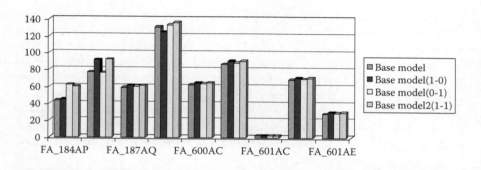

Figure 8.10   **Average daily inventory level of front axles.**

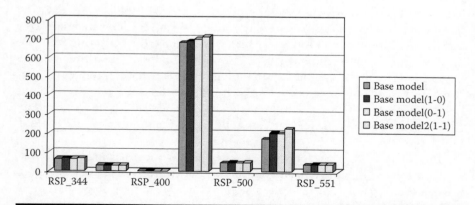

Figure 8.11   **Average daily inventory level of rear spring coils.**

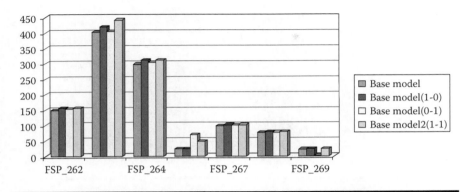

Figure 8.12   Average daily inventory level of front spring coils.

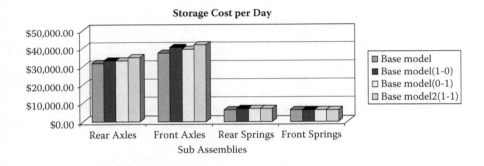

Figure 8.13   Average daily storage cost for all subassemblies.

The primary reason for the storage cost difference is the storage capacity of the bins. A storage bin can store only 10 axles or 192 spring coils. Therefore, when there is an increase in the average inventory level of the subassemblies, the number of bins corresponding to the axles increases drastically, but the number of bins corresponding to the springs increases marginally. This explains the considerable increase in storage cost of the axles compared to the springs. Furthermore, the consumption of subassemblies does not have a similar pattern. Among the subassemblies, the consumption rate of the individual models also makes a difference in the ability to handle product variety. For instance, among the front spring coils, there are a few spring models that are consumed at a very high rate, but the majority of the models have a low rate of consumption. This enables the front spring coils to absorb the added product variety with the corresponding result of no additional storage cost. A gradual increase in the inventory holding cost of the subassemblies is observed with the inclusion of new product variants (Figure 8.14).

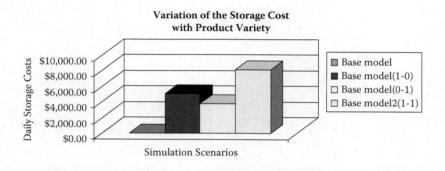

Figure 8.14   Increase in storage cost with added product variety.

8.3.6.2 *Monthly Production Sequence with Hourly Replenishment*

It was observed that with the hourly inventory replenishment policy, the average inventory level of the high-consumption subassemblies was low but that the average inventory level of the low-consumption subassemblies was high enough to prevent line stoppages. Thus, a new delivery policy was proposed and modeled. The high-consumption subassemblies are replenished on an hourly basis, and the low-consumption models are delivered at the beginning of every shift. Front Axle 601 AC, rear spring coils 345, 400, 500, and 551, and front spring coils 265 and 269 are replenished at the beginning of every shift, and the rest of the subassemblies are replenished hourly. The results are plotted in Figures 8.15 and 8.16.

The results show that average daily inventory holding cost and storage cost for the monthly schedule hourly replenishment model is less than the daily sequence hourly replenishment model. Theoretically, this is accompanied by an increase in material-handling cost, but this analysis is not included in the chapter. In summary, the proposed methodology provides insight into the behavior of the manufacturing system by capturing the variation of the inventory holding cost and the storage cost due to added product variety. Although it is expected that the inclusion of a new model will only increase the inventory level (and correspondingly inventory holding and storage costs) of the corresponding subassembly, the results of the analysis show that the inclusion of the new product variant also impacts the inventory level of other subassemblies in the system. Thus, the final inventory level on the assembly line is determined primarily by the sequence of vehicle assembly and the delivery policy. The simulation model successfully captures these parameters. Hence, the model is an analytical tool for production managers to make informed decisions regarding a new product introduction. The model can also be used to study the inventory holding and storage costs for alternative replenishment policies. The result from the developed methodology suggests that product variety in a mixed-model assembly line can be handled successfully by altering the assembly sequence and the

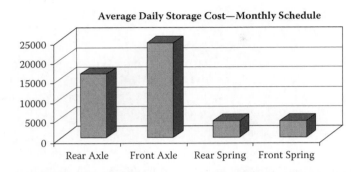

Figure 8.15   Average daily inventory storage cost for all subassemblies.

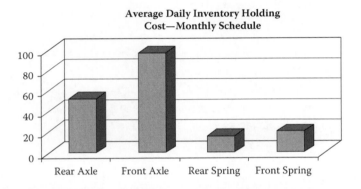

Figure 8.16   Average daily inventory holding cost for all subassemblies.

delivery policy. The cost model is generated by capturing the model mix complexity in a PSG (product structure graph) and generating a schedule to feed the simulation model that runs on the JIT philosophy. The simulation model helps to reveal the impact of added product variety on inventory holding and storage cost. The cost model provides an analytical tool in manufacturing to estimate the projected increase in inventory holding and storage cost and also to study the impact of the manufacturing system on various material handling policies.

Authors

Ali K. Kamrani is an Associate Professor of Industrial Engineering. He is also Founding Director of the Design and Free Form Fabrication Laboratory at the University of Houston, USA. He received his BS in Electrical Engineering in 1984,

Master of Engineering in Electrical Engineering in 1985, Master of Engineering in Computer Science and Engineering Mathematics in 1987 and PhD in Industrial Engineering in 1991, all from the University of Louisville, Louisville, Kentucky. His research has been motivated by the fundamental application of systems engineering and its application in advanced design and development of complex systems. He is the Editor-in-Chief for the *International Journal of Collaborative Enterprise* and the *International Journal of Rapid Manufacturing*.

Arun Adat is the Supply Chain Strategy and Development Manager at Hewlett-Packard. He is responsible for developing worldwide product fulfillment strategies. Arun's expertise includes supply chain network design, emerging market assessment, new product introduction processes, and mathematical modeling of supply chains. Prior to joining Hewlett-Packard, Arun was the Logistics Solutions Analyst at Goodman Manufacturing. He designed and prototyped customized algorithms, demonstrated potential cost savings in freight and distribution, collaborated with IT, and implemented heuristics on the production system. Arun holds a Bachelor of Science in Mechanical Engineering from Manipal Institute of Technology, India, and a Master of Science in Industrial Engineering from University of Houston.

Maryam Azimi is a Software Project Manager at Lenovo Corporation. She received a PhD in industrial engineering from the University of Houston in 2011. Her research interests are systems engineering, data mining in health care, lean Six Sigma, and project management.

References

Aigbedo, H., and Y. Monden. 1996. A simulation analysis for two-level sequence scheduling for just-in-time (JIT) mixed-model assembly lines. *International Journal of Production Research* 34(11): 3107–24.

Akturk, M. S., and F. Erhun. 1999. An overview of design and operational issues of Kanban systems. *International Journal of Production Research* 37(17): 3859–81.

Anwar, F. M., and R. Nagi. 1997. Integrated lot-sizing and scheduling for just-in-time production of complex assemblies with finite set-ups. *International Journal of Production Research* 35(5): 1447–70.

Baykoc, O. F., and S. Erol. 1998. Simulation modeling and analysis of a JIT production system. *International Journal of Production Economics* 55(7): 203–12.

Benjaafar, S., J. S. Kim, and N. Vishwanadham. 2004. On the effect of product variety in production-inventory systems. *Annals of Operations Research* 126(1–4): 71–101.

Bukchin, J. 1998. A comparative study of performance measures for throughput of a mixed model assembly line in a JIT environment. *International Journal of Production Research* 36(10): 2669–85.

Carlson, G. J., and A. C. Yao. 1992. Mixed model assembly simulation. *International Journal of Production Economics* 26(4): 161–67.

Chu, C. H., and W. L. Shih. 1992. Simulation studies in JIT production. *International Journal of Production Research* 30(1): 2573–86.

Detty, R. B., and J. C. Yingling. 2000. Quantifying benefits of conversion to lean manufacturing with discrete event simulation: a case study. *International Journal of Production Research* 38(2): 429–45.

Dobson, G., and C. A. Yano. 2002. Product offering, pricing, and make-to-stock/make-to-order decisions with shared capacity. *Production and Operations Management* 11(3): 293–306.

Drexl, A., and A. Kimms. 2001. Sequencing JIT mixed-model assembly lines under station-load and part-usage constraints. *Management Science* 47(3): 480–91.

Garcia-Sabater, J. P. 2001. The problem of JIT dynamic sequencing. A model and a parametric procedure. *ORP3*, September 26–29.

Ishii, K., and M. V. Martin. 1996. Design for variety. A methodology for understanding the costs of product proliferation. *Proceedings of the 1996 ASME Design Engineering Technical Conferences and Computers in Engineering Conference*, August 18–22.

Li, J. W. 2003. Simulation based comparison of push and pull systems in a job-shop environment considering the context of JIT implementation. *International Journal of Production Research* 41(3): 427–47.

MacDuffie, J. P., K. Sethuraman, and M. L. Fisher. 1996. Product variety and manufacturing performance: Evidence from the international automotive assembly plant study. *Management Science* 42(3): 350–69.

Malik, S. A., and W. G. Sullivan. 1995. Impact of ABC information on product mix and costing decisions. *IEEE Transactions of Engineering Management* 42(2): 171–176.

Mane, N., S. Nahavanadi, and J. Zhang. 2002. Sequencing production on an assembly line using goal chasing and user defined algorithm. *Proceeding of the 2002 Winter Simulation Conference*.

Martinez, F. M., and L. M. A. Bedia. 2002. Modular simulation tool for modeling JIT manufacturing. *International Journal of Production Research* 40(7): 1529–47.

Mejabi, O., and S. Wasserman. 1992. Basic concepts of JIT modeling. *International Journal of Production Research* 30(1): 141–49.

Miltenburg, J. 1989. Level schedules for mixed-model assembly lines in just-in-time production systems. *Management Science* 35: 192–207.

Monden, Y. 1998. *Toyota Production Systems—An Integrated Approach to Just in Time*. London: Chapman & Hall.

Muralidhar, K., S. R. Swenseth, and R. L. Wilson. 1992. Describing processing time when simulating JIT environments. *International Journal of Production Research* 30(1): 1–11.

Ozbayrak, M., M. Akgun, and A. K. Turker. 2004. Activity based cost estimation in a push/pull advanced manufacturing system. *International Journal of Productions Economics* 87: 49–65.

Savsar, M. 1997. Simulation analysis of a pull-push system for an electronic assembly line. *International Journal of Production Economics* 51: 205–14.

Spedding, T. A., and G. Q. Sun, 1999. Application of discrete event simulation to the activity based costing of manufacturing systems. *International Journal of Production Economics* 58: 289–301.

Wang, D., and C. G. Xu. 1997. Hybrid push/pull production control strategy simulation and its applications. *Production Planning and Control* 8(2): 142–51.

Chapter 9

Optimizing Supply Chain Network Design

Mohammed Hussein Hassan and Haitham Abbas Ahmed Mahmoud

Mechanical Engineering Department, Helwan University, Cairo, Egypt

Contents

9.1 Introduction

The efficient design of a supply chain network (SCN) includes deciding and controlling the key variables affecting the activities of the supply chain (SC). There are three levels of decisions that must be optimized in the design of an SC: the strategic, the tactical, and the operational decisions. The strategic level's decisions develop main frames for the operation of the SC. Tactical decisions focus on considering the internal constraints of the organization in order to best utilize the available resources to maximize the total profit for the organization.

Third, operational decisions are the decisions that taken to ensure the removal of the obstacles faced by the organization in fulfilling the various customers' orders and, accordingly, achieving the objective of minimizing the total cost of the SC. One of the most important issues in designing the SC is the design of the logistics network, which is the main concern of this chapter.

The purpose of this chapter is to focus on the logistics issues, including the classification of the SC based on its logistics network. The optimization and mathematical modeling of a dynamic SCN are then developed. Solution methodologies used to solve the logistics network models are then presented. A hypothetical problem is also used to apply the mathematical model. The behaviors of dynamic SCs under different dynamic parameters change are then presented. Finally, a case study on dynamic supply chains from the ready-mixed concrete (RMC) industry is presented.

9.2 Types of Supply Chain Logistics Networks

The Council of Supply Chain Management Professionals (CSCMP) has defined the term *logistics* as "the part of supply chain management that plans, implements, and controls the efficient, effective forward and reverse flow of materials of goods, services, and related information between the point of origin and the point of consumption in order to meet customers' requirements" (Council of Supply Chain Management Professionals 2011). A typical forward logistics network is shown in Figure 9.1.

Based on the status of the parameters of the network model, SC can be classified into static supply chains (SSCs) and dynamic supply chains (DSCs). An SC is said to be static if no changes occur in the characteristics or planning decisions in the chain logistics model over the planning horizon. Single-period planning SCs or even multiperiod SCs with unchanged parameters are considered as SSCs (Melo et al. 2005). The characteristics of an SSC comprise fixed facility locations, deterministic product demand, and stable working environment of the SC.

On the other hand, adding the state of dynamicity to the SC's logistics-formulated models is very important in order to approach the real-world conditions in which

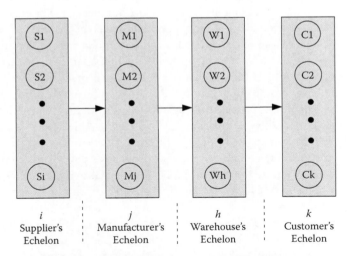

S: Supplier of raw materials, components, and subassemblies
M: Manufacturing facility
W: Warehouse
C: Customer zones or project sites

Figure 9.1 A typical forward logistics network.

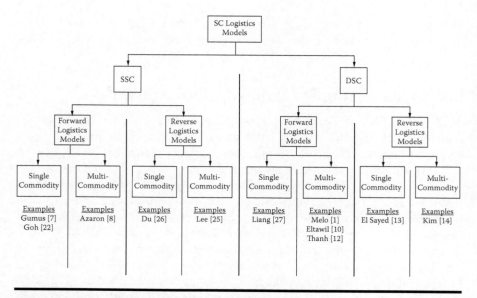

Figure 9.2 Classification of SC network models.

the SC works. The DSC is defined as "a supply chain in which one or more of the following characteristics could occur; dynamic relocation is required for the main facilities of the chain, increasing or decreasing the number of facilities over time in any echelon, changing variables of the supply chain, and the dynamic environment of the supply chain" (Hinojosa et al. 2008). A classification of the SC logistics design models with the aid of research work examples is exhibited in Figure 9.2.

From Figure 9.2, the SSC or the DSC can be classified according to the direction of the material movement through the chain into forward logistics chains or reverse logistics chain; the forward and the reverse logistics can be further classified as either single commodity or multicommodity. This classification is related to the number of products transported through the chain.

The forward multistage logistics network problem is defined as follows: given a set of facilities, including potential suppliers, potential manufacturing facilities, and distribution centers (DCs) with multiple possible configurations, and a set of customers with deterministic demands, the flow of material is in the downstream direction. The functions of the chain are the procurement of raw materials, the conversion of these materials into intermediate and finished products, and the distribution of these products to the retailers and customers in order to minimize the total cost of the chain (Goetschalckx et al. 2000).

According to Alshamarani et al. (2007), the reverse SC can be defined as "a chain in which the materials flow is in the upstream direction where the product over demanded or unsatisfied by the customer is returned back to the previous echelons of the chain for the purpose of remanufacturing, reusing, and recycling."

The reverse logistics networks are classified according to the network structure into two types: the recovery network that is fully focusing on the recovery activities and the closed-loop network that is integrated with the forward network (Melo et al. 2009).

9.3 Mathematical Modeling of DSC Models

The mixed-integer linear programming (MILP) is the most widely used method for formulating the logistics networks mathematically. Most of the developed models in this field are single objective with various types of dynamic SC problem's constraints. Two types of decision variables are usually used: the binary decision variables for the facility locations decisions and the integer decision variables for the decisions of the distributed quantities of products and materials.

9.3.1 Objective Function of the Mathematical Model

The objective of the model is to minimize the total SC planning cost over the whole planning horizon.

The total cost function of the SC comprises the following costs:

- Materials costs
- Materials transportation cost (from suppliers to manufacturers)
- Products manufacturing cost
- Products transportation cost (from plants to warehouses)
- Products transportation cost (from warehouses to customers)
- Materials holding cost (in the plants)
- Products holding cost (at warehouses)
- Products shortage cost
- Non-utilized capacity costs
- Fixed costs

Examples of the cost items included in the total SC costs function are as follows.

Materials Costs:

$$\sum_{\forall i}\sum_{\forall j}\sum_{\forall m}\sum_{\forall t}[(MPC_t^m + MTC_{ij,t}^m D_{ij})Q_{ij,t}^m + (MHC_{jt}^m HQ_{jt}^m)]S_{it}^m M_{jt}^p \qquad (9.1)$$

These costs include the materials purchasing, transportation, and holding costs. First, the purchasing costs equal the number of units from (Q) each material m transmitted from supplier i to plant j multiplied by the unit price of this material (MPC). Second, the transportation costs are the product of the transportation cost per unit product per unit distance (MTC) and the distance (D). Third, the holding costs are calculated by multiplying the holding cost per unit (MHC) by the total quantity held in stock (HQ). All of the three cost types are multiplied by binary decision variables to indicate whether the supplier S and plant M are active at this time period (t).

Products Costs:

$$\sum_{\forall j}\sum_{\forall h}\sum_{\forall k}\sum_{\forall p}\sum_{\forall t}[(PMC_t^p + PTC_{jht}^p D_{jh})P_{jht}^p + (PTC_{hkt}^p D_{hk})P_{hkt}^p$$

$$+ (HC_{ht}^p HP_{ht}^p)]M_{jt}^p W_{ht}^p C_{kt}^p$$

(9.2)

The cost items constituting the total products costs are the products manufacturing costs, the products transportation costs to both the warehouse sites, and the customer zones, and the products holding costs, where the symbols used are as follows:

> j, h, and k are the manufacturer, warehouse, and the customer indices, respectively. p is used as the product type index. The cost coefficients used in the equation are PMC, PTC, and HC for the product manufacturing cost, the product transportation costs, and the product holding costs in the same order. The location-allocation binary decision variables are M, W, and C for manufacturers, warehouses, and customers. The integer decision variables are P for the transported quantities of products and HP for the quantities of products held in stock.

9.3.2 Mathematical Model Constraints

Several types of constraints are included in the model in order to approach the real working environment of an SC. The demand satisfaction constraints and the capacity constraints are examples of the constraints handled by the model. Examples of the vital constraints included in DSC network models are as follows.

9.3.2.1 Demand Satisfaction Constraints

Materials Demand:

$$\sum_{\forall i}(Q_{ij,t}^m \cdot S_{it} + HQ_{j,t}^m)M_{jt}^p \geq (P_{jkt}^p + HQ_{jt}^m)M_{jt}^p \qquad \forall j, \forall m, \forall t$$

(9.3)

This constraint ensures that each manufacturing facility receives enough material or component quantity necessary for production. Demand satisfaction is realized from the materials purchased from suppliers in the current period and from the material stored from previous periods in the manufacturing facility.

Final Products Demand:

$$\sum_{\forall h} P_{hkt}^{p}\, W_{ht} \geq \beta_{kt}^{p}\, PD_{kt}^{p}\, C_{kt}^{p} \qquad \forall k,\forall p,\forall t \qquad (9.4)$$

This constraint ensures the demand satisfaction of the customer from the final product stored in the warehouses. All the customers' demand (*PD*) is satisfied from the produced products within service level β_{kt}^{p} specified for each customer from each product in each time period.

9.3.2.2 Capacity Constraints

The general form of the capacity constraints is that the sum of all materials and products quantities transferred from a selected origin node to all the selected destination nodes does not exceed the origin node's capacity from these materials or products at any time period *t*. This type of constraint is used for both the materials suppliers' capacity and the products manufacturers' capacity.

Materials Suppliers' Capacity:

$$\sum_{\forall j} Q_{ijt}^{m}\, S_{it}^{m}\, M_{jt}^{p} \leq Cap_{it}^{m}\, S_{it}^{m} \qquad \forall i,\forall m,\forall t \qquad (9.5)$$

where Cap_{it}^{m} is the material capacity of supplier *i* from material *m* at time period *t*.

Products Manufacturers' Capacity:

$$\sum_{\forall h} P_{jht}^{p}\, M_{jt}^{p} W_{ht}^{p} \leq Cap_{jt}^{p}\, M_{jt}^{p} \qquad \forall j,\forall p,\forall t \qquad (9.6)$$

where Cap_{jt}^{p} is the product capacity of manufacturer *j* from product *p* at time period *t*.

9.3.2.3 Budgeting Constraint

(materials purchasing cost
+ materials transportation cost
+ products manufacturing cost
+ products transportation to warehouses cost
+ products transportation to customers cost
+ materials holding cost + products holding cost
+ shortage cost + non-utilized capacity cost + fixed costs) $_{t} \leq B_{t} \forall t$ (9.7)

This constraint is intended to limit the total cost of the SC at any time period *t* within a planned budget B_t set by the planning management at this time period.

9.4 Solution Methodologies Used to Solve the SCN Models

A detailed classification of the used solution methodologies in the reviewed articles is exhibited in Figure 9.3. In general, two approaches are used for solving the developed models:

- Using a commercial solver in an attempt to find the exact solution to the problem tackled by the SCN model.
- Relying on the search heuristic techniques as an alternative to the commercial solvers. These heuristics are characterized by their flexibility to deal with different models with different sizes.

9.4.1 Commercial Optimization Packages Used for Solving the SCN Models

These packages are mainly based on the branch-and-bound algorithm and usually give optimal solutions to the optimization models. The main packages used for the optimization purpose are as follows:

- The LINGO® (versions 8.0, 10, 11, and 12), created by the Lindo Systems Incorporation, were used to solve the models designed by Melachrinoudis

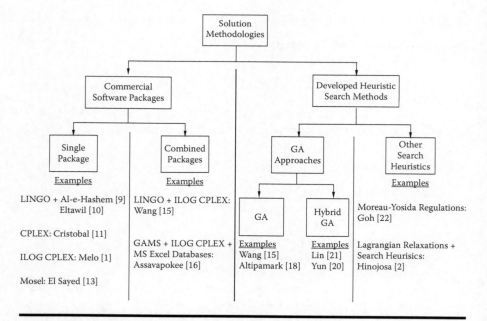

Figure 9.3   Classification of the solution methodologies.

and Min (2007), Gumus et al. (2009), Azaron et al. (2008), Al-e-Hashem et al. (2011), and Eltawil et al. (2007).

■ The ILOG CPLEX® (versions 9.1 and 10), developed by IBM, were used to solve the models designed by Cristobal et al. (2009) and Melo et al. (2005).

■ Other optimization packages include the Xpress-MP solver (used by Thanah et al. (2008)), the Mosel (used by El-Sayed et al. (2010)), and the OPL Studio (used by Kim et al. (2006)).

■ In some research work, more than one optimization package is used in solving the design models. For example, Wang and Hsu (2010) used the LINGO and the CPLEX. Also, Assaavapokee (2012) used the GAMS, the CIPLEX, and the MS Excel.

9.4.2 Search Heuristic Techniques Used for Solving the SCN Models

In much of the reviewed research work, search heuristic techniques were developed for these researches such as Wang and Shu (2007), Altiparmark et al. (2006), and Sourirajan et al. (2009). In other researches, other tools are combined with the genetic algoritm (GA) to form the hybrid genetic algorithm (hGA) tools. For example, Yun et al. (2009) supplement the GA with an adaptive local search method. Also, Lin et al. (2009) used a hybrid evolutionary algorithm that is composed of the GA, the Wagner–Whitin Algorithm, and a fuzzy logic controller.

On other hand, another category of search heuristics was used that are customized for the designed model. Examples include the Moreau–Yosida regulations, which are mainly used to solve the nonlinear convex models such as that designed by Goh et al. (2007). The surrogate constraint method is updated to solve the model established by Bidhandi et al. (2009). In addition, Lagrangian relaxations were used by Hinojosa et al. (2008), and then, a local search heuristic was used to solve their model.

9.5 Example of Modeling and Solving Dynamic Supply Chains Problem

The purpose of this section is to model and solve a hypothetical SC problem that contains the important dynamic features existing in the real-world SC problems. The problem is a two-period multicommodity SC; the proposed SC logistics network is shown in Figure 9.4. The problem has two scenarios for the two time periods pertaining to the number of nodes in each echelon of the chain and the quantities of materials and products transported through the chain. The scenarios, the generated cost parameters, and the formulation of the model are detailed in the following sections. Then, the solution methodology for the problem and the developed solutions will be set forth.

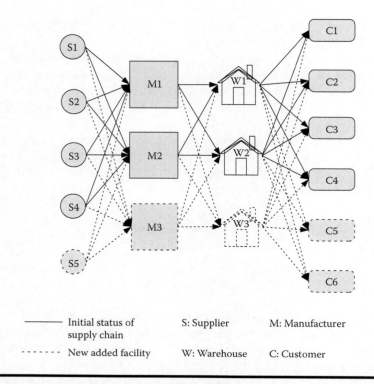

Initial status of supply chain ——————

New added facility - - - - - -

S: Supplier
W: Warehouse

M: Manufacturer
C: Customer

Figure 9.4　Logistics network for the test problem.

9.5.1 Logistics Network Scenario for Time Period 1

The following conditions are used when modeling the logistics network mathematically:

- There are *four* customer areas that require satisfaction for their demands of end products. The chain produces *three* products that can be ordered by customers.
- The SC has *two* warehouses that can supply the three products to the customers.
- There are *two* manufacturing plants in service. Each plant has the ability to produce and to be assigned to any of the three products.
- The *three* products are produced using *five* different materials and components.
- *Four* candidate suppliers are available to supply the required raw materials according to the suppliers' capacities of the different raw materials.
- Storage of the raw materials is allowed at the manufacturing facilities, while storage of the end products is allowed at the warehouses.
- The end products' demands are known and deterministic in each time period, and also the cost changes between nodes of the SC are normally distributed.

9.5.2 Logistics Network Scenario for Time Period 2

Several changes have occurred in the number and sites of the nodes in each echelon of the chain. The main changes can be stated as follows:

■ The customers' locations that require to be served are increased to *six*. Two new products are included in the set of products supplied by the chain so that the number of products becomes *five*.

■ As a result of the increase of customers' demands for end products, the number of warehouses should be increased by adding an extra warehouse. In period two, the number of existing warehouses is *two*, and there are two candidate warehouses' locations to select to open one of them.

■ The manufacturing plants will also be increased to be *three* after selecting to open a new plant. There are two potential plant' locations to select for this new plant.

■ The *five* products are produced using *eight* raw materials and components.

■ The suppliers' list includes one new supplier, so the suppliers' echelon has five nodes.

9.5.3 Numerical Representation for the Test Problem

The problem is represented numerically through the random generation of the input data parameters including the coefficients of the objective function and the initial conditions of the SC, too. Many different parameters and coefficients were generated randomly using the package of *Minitab16*.

The number of variables of the mathematical model representing the problem is 160 variables (50 for period 1 and 110 for period 2) with 164 constraints (57 constraints for period 1 and 107 constraints for period 2). The problem is considered as *NP* hard with (35×10^6) solutions.

9.5.4 Test Problem Solution

The solution methodology used for solving the current test problem is the *hybrid genetic algorithm* (*hGA*). It is composed of two stages: the *genetic algorithm* (GA) and the *pattern search* (PS) optimization. The two stages are sequential as shown in Figure 9.5 where the *GA* provides a reasonable initial feasible solution point

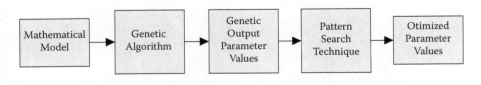

Figure 9.5 hGA solution methodology stages.

that can be used for further search and optimization by the PS to develop the final solution point. The *Matlab R2010a* package was used for the optimization process by using the *Global Optimization* toolbox.

9.5.4.1 GA Search Stage

GAs (as stated by Gen and Lin (2006)) are stochastic search algorithms that mimic the mechanism of natural selection and natural genetics. GAs start with an initial set of random solutions called the *population*. Each individual in the population is called a *chromosome*. To create the next generation, the elitist individuals survive and are selected to the next generation. In addition, new chromosomes, called *offspring*, are formed by either merging two chromosomes from the current generation using a crossover operator or modifying a chromosome using a mutation operator. Evolutionary algorithms operate on a population of potential solutions applying the principle of survival of the fittest to produce better and better approximations to a solution. Figure 9.6 shows the procedure of the GA in a flowchart. The proposed GA is designed in terms of optimizing the main operations and operator parameters of the GA. The main optimized operations of the proposed GA are

1. Problem and variable representation
2. Selection operation
3. Crossover operation
4. Mutation operation
5. Using one of the population models (migration model)

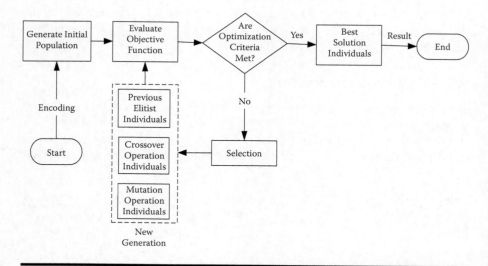

Figure 9.6   Flowchart of the main stages of the GA search technique.

Table 9.1   GA Initialization and Procedure

Procedure: GA Initial Steps	Procedure: GA Operators
Initialization	**Crossover operator**
Model representation stage	Use the intermediate crossover algorithm
Set f (objective function) as	
$f = Min \sum_{\forall i} C_i X_i$	Set the ratio R = 1
	Mutation operator
Set C (constraints) as $A_i X_i \leq b_i$	Use the uniform mutation method
Gene representation	Set the ratio R = 0.25(25% of the variables will be mutated)
Use real-valued representation in each chromosome	
Use 500 individuals as initial population	**Migration model**
	Use the migration method in both directions (forward and backward)
Chromosome selection	
Step 1. Fitness assignment	Set the migration fraction as 0.2 (ratio of interchanged individuals)
Use the shifted linear ranking fitness assignment method	Set the migration interval as 20 generations
Set the maximum survival rate as 2	
Step 2. Individuals selection	
Use the tournament selection method	
Set the tournament size as 4	
Set the number of elitist individuals as N1 = 50	
Set the number for crossover operations as N2 = 400	
Set the number for mutation operations as N3 = 50	

Table 9.1 shows the developed GA initialization process, the chromosome selection process, and the GA operators.

9.5.4.2 Pattern Search Stage

PS is an optimization algorithm that searches a set of points around the current solution point. This set of points, called a *mesh*, is formed by adding the current point to a scalar multiple of a set of vectors, called a *pattern*. In each step of the PS algorithm, a *poll* process is performed so as to decide whether there are points in

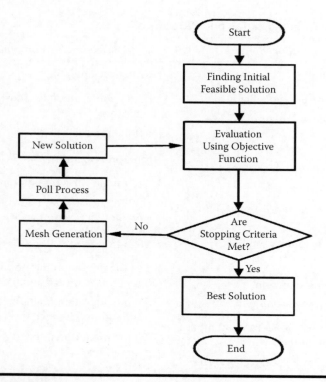

Figure 9.7 PS flowchart.

the mesh that improve the objective function's value at the current point. In the case of an objective function improvement, the new point becomes the current point at the next step of the algorithm, and a mesh expansion process is performed. Otherwise, the previous solution points are still the default solution point, and a mesh contraction process is performed. A flowchart for the used PS algorithm is drawn in Figure 9.7. The proposed PS procedure is shown in Table 9.2.

9.5.5 Test Problem Results

The solution of the developed multiperiod problem is represented in the values of the decision variables of the value of the fitness function and, further, the best values of this function were obtained by the two modules (GA and PS) and represented in Figures 9.8 and 9.9. The cost percentages shown in Figure 9.10 were calculated from the test problem's solution.

Decisions of other locations included in the solution are as follows:

▪ Open a new plant in location 1 of the two candidate locations.
▪ Open a new warehouse at the proposed location 1 of the two potential locations.

Table 9.2 The Used PS Procedure

PS Initialization	PS Algorithm
Set of inputs	**Step 1. Mesh generation**
• f (objective function) f $\in$ R^n	• Let $\{v_i\}$ be the pattern vectors
• *A, b* (the constraints' coefficients and the constraints' right-hand side) A, b $\in$ R^n	For i = 1:160
	$v_{i\in}\{-1, 0, 1\}$
• x_0 (start solution point), where $x_0 \in$ R^n	Generate the direction vectors $\{d_i\}$ by multiplying $\{v_i\}$ by a scalar Δ_i (mesh size = 1)
	Calculate $P_k: = \{x_k \pm d_i : i \in N\}$
	End
	Step 2. Poll Step
	For k = 0, 1, ...
	If $f(t) < f(x_k)$ for some t $\in$ P$_k$
	Set $x_{k+1} = x_t$
	Set $\Delta_{k+1} = 2*\Delta_k$ (Expansion)
	Otherwise
	Set $x_{k+1} = x_k$
	Set $\Delta_{k+1} = \Delta_k/2$ (Contraction)
	End

9.6 Analyzing the DSC Models under Different Levels of Dynamic Parameters

The purpose of these experiments is to show how the total SC cost per product unit (TSCU) is changed with a change in one or more of the SC dynamic parameters.

9.6.1 Design of Experiments for Study 1

In this study, the experimental work includes modeling of a dynamic four-echelon multiproduct SC's model with three time periods. The SC logistics scenarios for the three time periods comprise all the dynamic features considered in most of the practical SCs. In order to conduct the current study, two classes of factors are studied. Class 1 includes the main studied factors that affect directly the total cost of a supply chain. Factors involved in class 2 are the supply chain resources' factors that help in clarifying the effect of the main factors at different availability levels

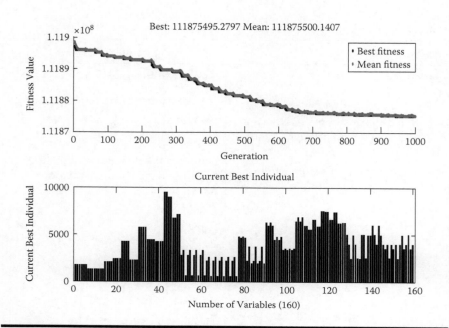

Figure 9.8   GA solution for the test problem.

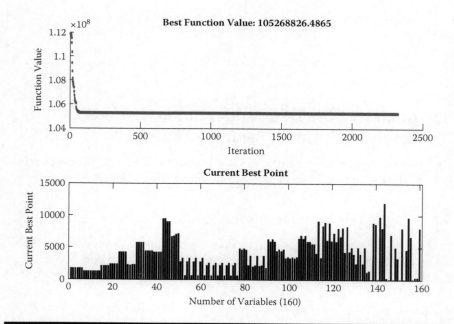

Figure 9.9   PS solution for the test problem.

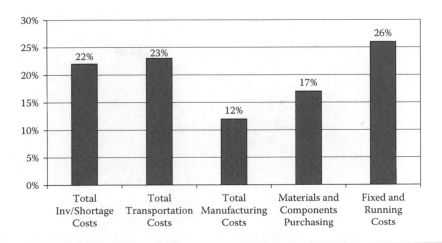

Figure 9.10   Percentages of the different cost items.

of resources. Factors of class 1 are selected on the basis of the used parameters in the relevant literature (Thanh et al.). These factors are

- Material purchasing cost per unit (MPC)
- Product manufacturing cost per unit (PMC)
- Product holding cost per unit per period (PHC)
- Products shortage cost per unit per period (PSC)
- Nonutilized capacity cost per unit per period (NUC)
- Product demand (PD)

On the other hand, the factors of Class 2 are considered on the basis of numerous experimentations performed in the current study. This class includes

- Material storage capacity (MSCap)
- Product manufacturing capacity (PMCap)
- Product storage capacity (PSCap)
- Availability of raw materials represented in the suppliers' capacities from each raw material (MCap)

In order to involve all the studied parameters with their levels in study 1, a number of SC scenarios must be built. The number of possible scenarios for study 1 is 182,250. The total number of problems solved, after fixing some of the factors in some problems, is 300 problems. The scenarios built are categorized into groups of problems, and each group is concerned with one of the studied parameters. These groups of problems were designed and optimized; the designed groups are as follows.

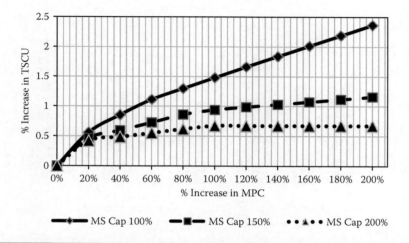

Figure 9.11   Effect of increasing MPC on the TSCU.

9.6.2 Effect of the Material Purchasing Cost

The effect of changing the MPC for the different used raw materials on the TSCU is shown in Figure 9.11. In general, it can be observed that the TSCU increases directly with the increase in the MPC. The change of the increasing pattern is influenced by three factors that are

■ The price of the material itself
■ The number of periods the material is utilized and purchased
■ The assigned storage capacity for each raw material

As shown in Figure 9.11, the relationship between the MPC increasing percentage and the TSCU is absolutely linear. This can be explained as follows:

■ The increase in the TSCU possibly occurs due to either the increase in the MPC or the increase in the total material holding cost.
■ With further increase in the MPC, it is economically beneficial to decide to use a larger proportion of the material storage capacity; the rate of increase in the TSC becomes smaller.
■ This further explains that increasing the assigned storage capacity by 50% and 100% has a slight negative effect on the TSCU for different values of the MPC.

9.6.3 Effect of the Product Manufacturing Cost

The effect of PMC increase on the TSCU is represented in Figure 9.12. Generally, the TSCU increases linearly with the increase in the PMC. There are two main

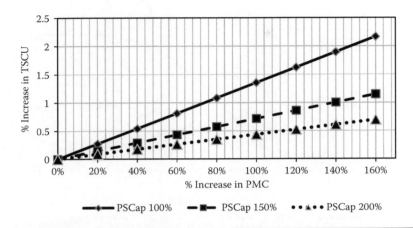

Figure 9.12   Effect of increasing PMC on the TSCU.

factors combined with the PMC that influence the slope of the plotted straight lines:

- The difference in the values of the PMC for the different products
- The manufacturing and storage capacities allowed for each product

Increasing the production and storage capacities for the products leads to the increase in the service levels associated with these products, causing considerable decreases in the total shortage costs. Also, higher storage capacities will allow the supply chain to store quantities of products as much as possible to reduce the manufactured quantities in the next periods with higher costs. That is why the TSCU decreases with an increase in the product storage capacity.

9.6.4 Effect of the Product Holding Cost

Figure 9.13 shows the effect of increasing the PHC on the TSCU. Results show that the increase in PHC causes the TSCU to be increased until a certain TSCU value is reached, and then the cost remains the same with further increases in the PHC. The behavior of the TSCU increase with increasing the PHC can be explained as follows:

- At small values of increase in the PHC, the TSC increases because of the increase in the total holding costs.
- With a further increase in the PHC, the extra production decision replaces completely the holding decisions. This is why the TSCU remains constant with higher increases in the PHC.

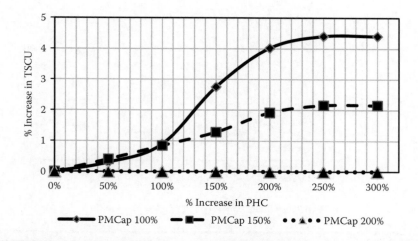

Figure 9.13   Effect of increasing PHC on the TSCU.

- As the PMCap level increases (150% and 200% of the original capacity), the rate of increase in the TSCU is considerably reduced. The reason for this behavior is that higher PMCap levels allow producing higher proportions of the demanded quantities and reducing the dependence on the storing option.

9.6.5 *Effect of the Product Shortage Cost*

Figure 9.14 shows the relation between the PSC increase and the relevant increase in the TSCU. The increase in the PSC makes the TSCU increases until it reaches a point of stability and remains constant with further increases in the PSC. This can be explained in the following points:

- In the case of small values of increase in the PSC, it is more economical to have a shortage in products and pay the penalty costs instead of producing and keeping in stock.
- In the case of higher levels of increase in the PSC, it is more economical to fulfill many of the customer orders rather than having shortages. The fulfillment of customer orders increases by increasing the PSC, resulting in the gradual elimination of the shortage option. After reaching 100% fulfillment of customer orders (due to increase in PSC), the percentage increase in TSCU remains constant even with an increase in PSC.
- The raw materials' capacities of the suppliers play a vital role in decreasing the percentage of increase in the TSCU because the availability of raw materials decreases the total shortage quantities. Hence, the total cost of the SC is greatly reduced.

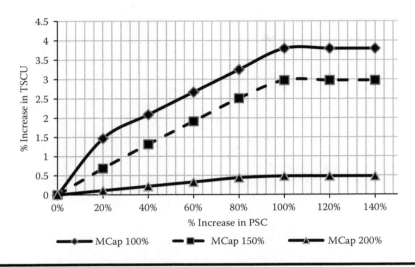

Figure 9.14   Effect of increasing PSC on the TSCU.

9.6.6 *Effect of Product Nonutilized Capacity Cost*

The effect of the NUC on the TSCU is shown in Figure 9.15. It can be noticed that the TSCU increases linearly with the percentage increase in the NUC. The behavior of the TSCU (the linear increase) with increasing the NUC can be related to the high dependence on the production and storage option with its high cost. Hence, the service level of the product is greatly increased. Even though the total shortage cost will be greatly reduced with increasing NUC, the extra production causes increase in the raw materials costs, the production costs, and the holding costs. These costs are not balanced by the reduction in the PSC. Increasing the production and storage capacities leads to an increase in the nonutilized capacity and, in turn, an increase in the TSCU.

9.6.7 *Effect of Product Demand*

As a factor, the effect of increasing the PD on the TSCU is shown in Figure 9.16. Generally, there is a linear increase in the TSCU on account of the PD increase. The increase in the total cost of the SC is due to the increased quantities produced from the products with increased demand. Also, the total shortage costs are greatly increased with PD increase, resulting in higher TSCU. Increasing the production and storage capacities decreases the slope of the plotted straight lines (i.e., decrease the range of increase in TSCU); increasing these capacities helps in reducing the shortage quantities, and accordingly the shortage costs and the TSCU are reduced.

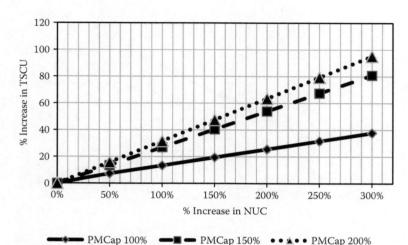

Figure 9.15   **Effect of increasing NUC on the TSCU.**

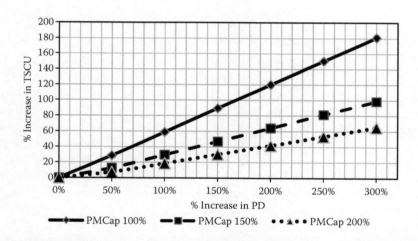

Figure 9.16   **Effect of increasing PD on the TSCU.**

9.7 Case Study on Dynamic Supply Chains: Supply Chain of the Ready-Mixed Concrete Industry in Egypt

Ready-mixed concrete (RMC) is a specialized multicomponent product in which a mixture of cement, water, aggregates comprising sand and gravel or crushed stone, and other ingredients are weigh-batched at a plant in a central mixer or truck mixer, before delivery to the construction site in a plastic or unhardened state ready for placing by the builders.

Advantages of using the RMC include elimination of storage space for basic materials at building sites, the production of a better-quality concrete, the great reduction of building time, and the reduction of building cost. On the other hand, the shortcoming of using a predetermined concrete mixture is the criticality of the traveling time from the centralized mixing plants to the construction site (ready-mix should be placed within 90 min of batching at the plant).

9.7.1 Supply Chain of the RMC Company

An SCN for a typical RMC company is shown in Figure 9.17. It contains the three main traditional echelons of an SC (suppliers; manufacturers, which are the RMC mixing plants; and customers, which represent the sites of the construction projects), in addition to transportation to demand sources. RMC can be produced in different mixing locations, mixing plants (MPs). The produced concrete may be categorized into two major types: the *regular* type and the *water-resistive* type. These two types can in turn be categorized into numerous products and mixtures; each product has its specific additives according to the type of construction activity

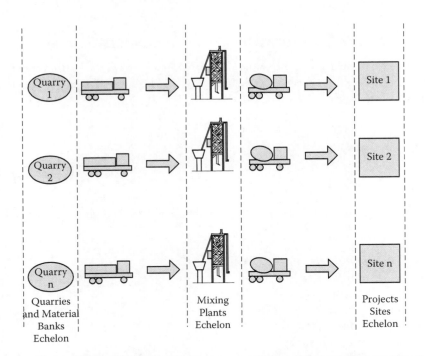

Figure 9.17 RMC supply chain echelons.

in which the concrete will be used and the required strength of the concrete. Since the RMC product has a very limited shelf life, all the quantities of the produced concrete are transported directly to the locations of the projects. The mobile concrete mixers are used for this function to prevent the concrete from solidifying. At the construction site, concrete pumps, either stationary or movable, are used for feeding the concrete to the specific locations of work. Obviously, demand sources (on-site projects) for RMC products are different, and each may be characterized by its dynamic pattern of demanded quantities.

9.7.2 Equipment and Mechanization of the RMC Processes

The RMC main processes and the required equipments for each process are as follows:

- The mixing operation, either in the central mixing plants or the movable plants, uses the production utilities, which are the loaders, the electrical generators, and the cooling chillers.
- The truck mixers transport the mix from the plant to the construction sites.
- Feeding the mix in the site is done using the (either fixed or movable) pouring pumps.

9.7.3 The RMC Sector of the Arab Contractors Company

The projects of the Arab Contractors Company (ACC, one of the leading construction companies in Egypt) in year 2005/2006 can be divided according to their locations into four groups:

- Cairo and Giza projects
- Dahshour projects
- Delta projects
- Upper Egypt projects

The construction projects served by the RMC sector are 12, which are located as follows:

- Gawharet Haras Gomhory, Heliopolis, Cairo.
- Faysal tunnel, Giza.
- Athar Elmabna port, Giza.
- South Dahshour bridge, Dahshour.
- Tanta bridges, Tanta.
- Fixing unit, Dahshour.
- Seket Elwaily bridge, Ramses, Cairo.
- Foundation unit, Dahshour.

Table 9.3 Demanded Quantities of RMC in Year 2005

	Demand in m³ for Each Month					
Project Site	*7/2005*	*8*	*9*	*10*	*11*	*12*
Gawharet Haras Gomhory	3,005	3,384	3,965	3,937	4,048	4,967
Faysal tunnel	450	1,000	1,000	950	700	—
Athar Elmabna port	150	250	320	400	500	600
South Dahshour bridge	2,100	2,300	2,000	2,500	2,400	2,700
Hawamdeya bridge	—	50	100	300	400	650
Hawees Esna	—	9,830	8,720	5,800	3,900	—
Rayaah Abbasy	—	110	688	693	1,370	5,465
Tanta bridges	500	1,200	1,250	500	1250	500
Fixing unit	120	2,160	1,620	1,270	220	934
Lebnan square bridge	—	—	1,700	1,650	900	1,500
Seket Waily bridge	1,000	1,000	1,150	1,100	600	1,000
Foundation unit	3,500	3,500	3,500	3,500	3,500	3,500
Total	10,825	24,684	26,013	22,600	19,788	20,856

- Hawamdeya bridge, Giza.
- Hawees Esna, Esna.
- Rayaah Abbasy, Zefta.
- Lebnan square bridge, Giza.

The demanded quantities of the RMC products for year 2005 (from July to December) are shown in Table 9.3.

9.7.4 Mathematical Modeling of the RMC Supply Chain

Based on the observation of the RMC process sequence of the Arab Contractors Company (ACC), one of the leading construction companies in Egypt, the generic DSC's optimization model developed in Chapter 3 is modified in order to fit the SC of the RMC industry. The basic features added to the original model in order to fit Case Study 1 are as follows:

- The main decision variables concerning the delivered quantities of raw materials and the delivered quantities of RMC are continuous. Hence, the solution time of the problem is considerably large. This feature is covered by using a continuous variable in the cost function and the constraints' equations.

■ Important binary decision variables are also added to consider the assignment process of the equipment owned by the company (i.e., the moving plants, the generators, the loaders, the pumps, and the truck mixers) or to depend on the leasing option of the equipment.

■ Special constraints are involved in the model; the added constraints comprise the limited number of equipment of each type owned by the company. The constraint of the RMC's shelf life is also contained in the model.

The problem is to make the best use of the stationary plants and to rely mainly on this type because of their low operating and production costs. In case of increased product demand, the movable plants will help in fulfilling the excess project demands of the RMC.

As an objective, it is intended to make a multiperiod production and transportation plan for the RMC sector of the company. This plan includes the crucial decisions of the logistics network: the plant locations' decisions assignment process for the projects' sites, the quantities of raw materials and final products transported through the network, and assignment decisions of the different equipment (given that the planning horizon is 6 months).

9.7.5 Cost Basis for Calculating the Cost Coefficients

The bases for calculating the different cost parameters of the RMC (according to the ACC's policy) are as follows:

■ All the mixing plants and other supplements have exceeded their useful life. The annual depreciation charge and the annual insurance are calculated according to the current market values of the equipment as follows:

■ The equipment annual depreciation is 40% of its current value. Further, the annual insurance of the equipment is 2.5% of the current market value.

■ The workforce required to operate a mixing plant is an engineer, an operating worker, five workers, electrical technician, mechanical technician, and one control electrician. The total labor cost is L.E. 11,200 monthly.

■ Other operating or leasing costs are recorded in Table 9.4.

9.7.6 Shutdown Maintenance Schedule

All the RMC plants (fixed and movable) have exceeded their useful life. For this purpose, a shutdown maintenance plan that is given in Table 9.5 is adopted in order to allow for the necessary maintenance activities. This plan takes into consideration

Table 9.4 Equipment Operating and Leasing Costs

Equipment	Leasing and Operating Costs (L.E./day)
Movable plant	1300
150 kVA generator	450
250 kVA generator	400
Loader	300
Movable pump (18 m³)	600
Movable pump (28 m³)	950
Movable pump (36 m³)	1350
Fixed pump	600
Truck mixer	527

the RMC demand volume for each project, the plants' production costs, and the plant's physical status.

9.7.7 Solution of Case Study 1 Model

The mathematical equations are represented using the cost data indicated in the preceding tables. Excel spreadsheets were designed to calculate all the cost coefficients of the objective function, and the constraints' coefficients and the constraints' right-hand side. The following guides are used in solving the model:

- Develop a set of supply options of RMC for each construction site (e.g., using stationary plants 1, 2, 3, or from a movable plant for site 1).
- Exclude any plant from the list that is distant from the construction site longer than 60 km (as the product's shelf life constraint will not be realized).
- Enter the created mathematical equations for the construction sites and their available supply options and then solve the multiperiod model.

The Lingo 10 optimization package is used to solve the model and to determine the optimized production and transportation plan concerning the assignment decisions of the available resources and the plants that serve each construction site. In addition, the optimized distributed quantities of the RMC are included in the optimal solution. An example of the optimized quantities of transported RMC from the mixing plants to the construction sites and the optimized distribution routes of the network are exhibited in Figure 9.18.

The total cost of the optimized supply chain and its distribution for the 6 months is shown in Table 9.6.

Table 9.5 Shutdown Maintenance Schedule for the RMC Plants

Central Mixing Plant	Location	Capacity (m^3/month)	RMC Quantities (m^3/month)					
			Month 1	Month 2	Month 3	Month 4	Month 5	Month 6
Kabag	Dahsh.	4186	on	on	off	on	on	off
Liebherr	Dahsh.	3151	off	on	on	off	on	on
Stetter	Dahsh.	2898	on	off	on	on	off	on
Granier	Esna	9016	off	on	on	off	on	on
Liebherr	Esna	3151	off	on	on	on	off	on
Elba 105	Nasr City	9016	on	on	off	on	on	off
Arabau	Zefta	4186	on	on	off	on	on	off
Liebherr	Moneeb	2898	on	off	on	on	off	on

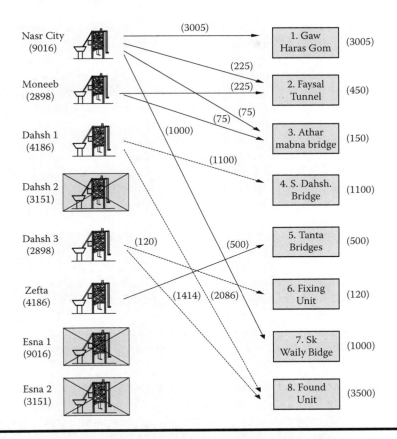

Figure 9.18 Optimized SC network for time period 1 (July 2005).

Table 9.6 The Optimized SC's Total Cost and Detailed Costs

Month	1	2	3	4	5	6	Total
Cost (L.E.)	3,062,169	6,660,370	7,093,644	5,836,475	5,224,070	5,344,927	33,181,191

9.8 Conclusion

Optimizing the design of an SCN, especially that existing in a dynamic environment, is a very important issue. It helps solving many problems inherent in the continuous changes in the SC parameters. This chapter addresses the issue by providing a classification of the SCN models and the solution methodologies. The mathematical formulation of the model is discussed via the MILP.

A hypothetical test problem is used to apply the developed dynamic model where an *hGA* comprising the GA and the PS optimization methods are used for solving the problem.

An experimental study is used to verify the model under different levels of the dynamic parameters. An SC of the RMC products is used to validate the DSC model, and the results assert that the dynamic model suits real SC problems existed in dynamic working conditions.

Now our attention turns to the design of risk-based SC models. This will help in tackling the uncertainties concerned with the dramatic changes in the prices of raw materials, the availability of different raw materials, and the availability of different resources. These situations represent common problems existed in the Egyptian market and are an integral part of executing the on-site projects there.

Authors

Mohammed Hussein Hassan is a Professor of Industrial Engineering at the Faculty of Engineering, Helwan University, Egypt. His research interests include Industrial Engineering, Operations Management, Machine Scheduling, Supply Chain Management, Quality Inspection Systems, Cellular Manufacturing System, Agile/Lean Manufacturing Systems, with an emphasis on using various optimization algorithms.

Haitham Abbas Ahmed Mahmoud is an Assistant Professor of Industrial Engineering at the Faculty of Engineering, Helwan University, Egypt. His research interests include Supply Chain Management, Optimization and Operations Research, Lean Manufacturing Systems, with an emphasis on using the AI techniques such as neural networks and genetic algorithms.

References

Al-e-hashem, S., Malekly, H., and M. Aryanezhad. 2011. A multi-objective robust optimization model for multi-product multi-site aggregate production planning in a supply chain under uncertainty. *International Journal of Production Economics* 134(1): 28–42.

Alshamarani, A., Mathur, K., and Ballou, R. 2007. Reverse logistics: simultaneous design of delivery routes and returns strategies. *Computers and Operations Research* 34(2): 595–619.

Altiparmak, F., Genan, M., and Lin, L. 2006. A genetic algorithm approach for multi-objective optimization of supply chain networks. *Computers and Industrial Engineering* 51(1): 197–216.

Assavapokee, T., and Wongthatsanekorn, W. 2012. Reverse production system infrastructure design for electronic products in the state of Texas. *Computers and Industrial Engineering* 62(1): 129–40.

Azaron, A., Brown, K., Tarim, S., and Modarres, M. 2008. A multi-objective stochastic programming approach for supply chain design considering risk. *International Journal of Production Economics* 116(1): 129–38.

Bidhandi, H., Yusuff, R., Ahmad, M., and Abu Bakar, M. 2009. Development of a new approach for deterministic supply chain network design. *European Journal of Operational Research* 198(1): 121–28.

Council of Supply Chain Management Professionals (CSSMP). Accessed on November 16, 2011, http://cscmp.org/aboutcscmp/definitions.asp.

Cristobal, M., Escudero, L., and Monge, J. 2009. On stochastic dynamic programming for solving large-scale planning problems under uncertainty. *Computers and Operations Research* 36(8): 2418–28.

El-Sayed, M., El-Kharbotly, A., and Afia, N. 2010. A stochastic model for forward–reverse logistics network design under risk. *Computers and Industrial Engineering* 58(3): 423–31.

Eltawil, A., Elwany, H., and Megahed, A. 2007. Multi-commodity multi-period supply chain network design. *Proceedings of the 37th International Conference on Computers and Industrial Engineering* 182: 164–73.

Gen, M., and Lin, L. 2006. Genetic algorithms and their applications. In *Springer Handbook of Engineering Statistics*, ed. H. Pham, 749–73. New York: Springer-Verlag.

Goetschalckx, M., Vidal, C., and Dogan, K. 2000. Modeling and design of global logistics systems: A review of integrated strategic and tactical models and design algorithms. *European Journal of Operational Research* 143(1): 1–18.

Goh, M., Lim, J., and Meng, F. 2007. A stochastic model for risk management in global supply chain networks. *European Journal of Operational Research* 182(1): 164–73.

Gumus, A., Guneri, A., and Keles, S. 2009. Supply chain network design using an integrated neuro-fuzzy and MILP approach: A comparative design study. *Expert Systems with Applications* 36(10): 12570–77.

Hinojosa, Y., Kalcsics, J., Nickel, S., Puerto, J., and Velten, S. 2008. Dynamic supply chain design with inventory. *Computers and Operations Research* 35(2): 373–91.

Kim, K., Song, I., Kim, J., and Jeong, B. 2006. Supply planning model for remanufacturing system in reverse logistics environment. *Computers and Industrial Engineering* 51(2): 279–87.

Lin, L., Gen, M., and Wang, X. 2009. Integrated multistage logistics network design by using hybrid evolutionary algorithm. *Computers and Industrial Engineering* 56(3): 854–73.

Melachrinoudis, E., and Min, H. 2007. Redesigning a warehouse network. *European Journal of Operational Research* 176(1) 210–29.

Melo, M., Nickel, S., and Saldanha, F. 2005. Dynamic multi-commodity capacitated facility location: A mathematical modeling framework for strategic supply chain planning. *Computers and Operations Research* 33(1): 181–208.

Melo, M., Nickel, S., and Saldanha, F. 2009. Facility location and supply chain management—A review. *European Journal of Operational Research* 196(2): 401–12.

Sourirajan, K., Ozsen, L., and Uzsoy, R. 2009. A genetic algorithm for a single product network design model with lead time and safety stock considerations. *European Journal of Operational Research* 197(2): 599–608.

Thanh, P., Bostel, N., and Peton, O. 2008. A dynamic model for facility location in the design of complex supply chains. *International Journal of Production Economics* 113(2): 678–93.

Wang, H., and Hsu, H. 2010. A closed-loop logistic model with a spanning-tree based genetic algorithm. *Computers and Operations Research* 37(2): 376–89.

Wang, J., and Shu, Y. 2007. A possibilistic decision model for new product supply chain design. *European Journal of Operational Research* 177(2): 1044–61.

Yun, Y., Moon, C., and Kim, D. 2009. Hybrid genetic algorithm with adaptive local search scheme for solving multistage-based supply chain problems. *Computers and Industrial Engineering* 56(3): 821–38.

Chapter 10

Shutdown Maintenance Scope of Work Assessment Model (SWAM): Model for Reducing Shutdown Maintenance Costs and the Loss of Production at Continuous Process Industries

Adel Al-Shayea

King Saud University, Department of Industrial Engineering, Riyadh, Saudi Arabia

Contents

Abstract

The aim of this chapter is to find a way to reduce the shutdown maintenance costs and the loss in production that is caused by performing shutdown maintenance activities. Following an assessment of the maintenance scope of work, the chapter proposes a model for reducing the shutdown maintenance costs and the loss in production without increasing the number of unplanned shutdowns. How the model is being applied in continuous process plants is presented and described as well in this chapter.

10.1 Introduction

In the industrial world, one of the challenging issues that face continuous process industries is the reduction of shutdown maintenance costs and the loss in the production that occurs as a result of performing shutdown maintenance activities while improving the availability of machines and equipment (Ghosh and Roy 2009, Gupta and Paisie 1997). In other words, the challenge is to increase profit while reducing the total maintenance cost.

This challenge stimulated the research in this chapter as well as the work in previous research to find methods to reduce the cost of shutdown maintenance and loss in production. Previous research can be classified into three categories. In the first and second categories, the approaches are aimed at reducing either the duration of shutdown maintenance or its frequency without increasing its cost in order to reduce the loss in the production, which in turn increases profit (Ashayeri et al. 1996, Baughman and Shah 1996, Chareonsuk et al. 1997, Koppelman and Faris 1994, Mathew and Rajendran 1993, Shuffelton 1998). In the third category, the approach is aimed at the reduction in the number of shutdown maintenance activities (scope of work of shutdown maintenance) that will reduce both the duration and the frequency of shutdown maintenance (Al-Shayea 2003). This third approach greatly contributes to the reduction of shutdown maintenance cost and the loss in the production that is caused by shutdown maintenance.

In this chapter, the discussion will focus on a model that is based on the positive relationship between the scope of work of shutdown maintenance and the shutdown maintenance cost, and the other positive relationship between the scope of work of shutdown maintenance and the loss of production (Al-Shayea 2003). In addition, the discussion in this chapter will include an explanation of how the model is used by means of a case study in a selected Saudi Arabian process plant.

10.2 General Characteristics of the Shutdown Maintenance Scope of Work Assessment Model (SMWAM)

The determination of the general characteristics of the shutdown maintenance scope of work assessment model (SMWAM) is an important step toward better understanding the model. It reveals important aspects such as the purpose, the type, and the use of model.

10.2.1 The Purpose of SMWAM

The purpose of SMWAM is to reduce the shutdown maintenance cost and loss in the production. The proposed way to accomplish this purpose is by transferring some activities from the list of shutdown maintenance activities to other in-process maintenance programs list of activities. The result of doing this is a reduction in the number of shutdown maintenance activities to be performed, which means a reduction in the amount of shutdown maintenance work, cost, and the loss in production that is caused by shutdown maintenance. At the same time, the unplanned shutdowns that could increase as a result of not carrying out these activities will not occur since these activities will be performed through other maintenance programs.

However, arbitrary selection of those shutdown maintenance activities that will be transferred could cause several problems. These problems are

- Increasing the overall maintenance cost because in some cases the cost of performing the transferred activities in other maintenance program is higher than the cost of performing them in the shutdown maintenance.
- Creating safety problems since some of those transferred activities need safety precautions, which can only be applicable, when the plant is in a shutdown situation.
- Creating operational problems because working with some of the transferred activities while the plant is running (not in shutdown situation) could cause full or partial unplanned shutdown.

The way to avoid such problems is by performing the process selection in a systematic manner. This means that the selection of the activities to be transferred should consider the three main factors that affect this process and which are related to the safety, operational, and economic issues. The reason for considering these three factors is because they are the only factors that have been identified by previous research (Al-Shayea 2003, Al-Najjar and Alsyouf 2003, Dekker and Scarf 1998, Smith and Hawkins 2004, Kelly 2006) as the important factors that affect the classification process of the critical activities. In other words, the selected shutdown maintenance activities to be transferred should not cause the following:

- Any harm or danger to the outside environment and to the workplace when they are performed by other maintenance programs (these activities are called *nonsafety critical activities*).
- Any partial or full unplanned shutdowns or any other operational issues such as the degrading of product quality when they are performed by other maintenance programs (these activities are called *nonoperational critical activities*).
- An increase in the overall maintenance cost when they are performed by other maintenance program in which the cost of performing these activities will be greater than performing them in the shutdown maintenance (these activities are called *noncost critical activities*).

10.2.2 Type of SMWAM

SMWAM can be classified according to its type in different ways. It can be classified as a mental and management science model since it uses quantitative and graphical forms to provide an assessment of the shutdown maintenance scope of work that in turn will help maintenance managers make decisions regarding the planning and the control of the shutdown maintenance. Also, it can be classified as a homomorphic model since it includes only those components that are relevant to the assessment of the shutdown maintenance scope of work. In addition, the model can be classified as a prescriptive model in the sense that it will suggest a modification to the existing shutdown maintenance scope of work assessment techniques. At the same time, it could act as a descriptive model since it aims to portray the relationship between the factors that affect the selection process and the activities. In general, the model is a planning and decision-making tool that can be used by maintenance professionals to reduce the shutdown maintenance cost and the loss of production through proper assessment of the shutdown maintenance scope of work.

10.2.3 Use of SMWAM

SMWAM is used to:

- Enhance the understanding of the reduction of the cost of shutdown maintenance and the loss of production.
- Stimulate creativity in the search for possible solutions to assess the proper shutdown maintenance scope of work that would end up with a reduction in the shutdown maintenance cost and in the loss in production.
- Evaluate alternative courses of action.
- Simplify the assessment process of the shutdown maintenance scope of work in order to clarify the relationships between the factors that affect the selection process of those activities to be included in the shutdown maintenance and the candidate activities.
- Examine the possible consequences of the different courses of action without exposing continuous process plants to cost, danger, or to consume their valuable time.

10.3 Modeling Method for Developing the Structure of SMWAM

The selected modeling method that is used to develop the structure of the model is a modified version of the separate preference elimination method (exclusionary screening), which is one of the discrete multicriterion (multiobjective) decision methods. This method treats each criterion separately and uses preference scales for each criterion to exclude or eliminate certain options (Goicoechea et al. 1982). The reason for using this method is that the problem at hand has three main criteria representing factors that are related to safety, operational, and economic issues, and it has a finite number of alternatives representing the two main courses of action that are the inclusion of an activity in the shutdown maintenance list or the exclusion of the activity from this list. These criteria and alternatives are equally important, which means that they are not vulnerable to pair-wise comparison techniques that in turn makes the selected method suitable for modeling this particular problem. By modifying and using this method, a great deal of structure and flexibility will be provided. Moreover, the use of the modified version of the method will provide transparency, safety, and clear audit trails, which means that it is very easy to trace the results of the decisions analyses if it is needed.

In addition to this method, the decision tree that shows the logical progression that occurs over time in terms of decisions and outcomes is used in order to develop a clear view of the structure of a problem and make it easier to determine the possible scenarios that can result if a particular course of action is chosen. The use of the decision tree is to ease the understanding of the problem and to enhance the communication process since the tree is a useful medium for communication.

10.4 Elements of SMWAM

SMWAM has three main elements. These elements are very important and represent the keystones for the selection process of the shutdown maintenance activities to be transferred (the evaluation process of shutdown maintenance activities).

The first element among the SMWAM model elements represents the decision options (alternatives) that make up the shutdown maintenance activities. One or a set of activities must be chosen to be transferred to other in-process maintenance programs. These alternatives (shutdown maintenance activities) will be evaluated (tested) against the three main factors that affect the selection process and which are related to safety, operational, and economic issues.

The second element of the SMWAM model represents the previously mentioned three main factors. These factors have been identified in the literature as critical factors. This, as it has been mentioned by Austin and Ghandforoush (1993) in the definition and description of the critical factor in the multicriterion decision model, means that if a shutdown maintenance activity does not meet the requirements of any one of these factors, it will be automatically eliminated from further consideration at the very beginning of the evaluation process, regardless of all other conditions that might exist. In particular, it means that failing to satisfy the requirements of any one of these factors during the evaluation of a certain shutdown maintenance activity will indicate that this activity has either a safety, operationally, or costly critical factor that will preclude its selection for transition to other in-process maintenance programs at the very beginning of the evaluation process, regardless of all other conditions that might exist. By safety, operationally, and costly critical activity, it is meant that this activity will create safety, operational, or economical problems if it isperformed by other in-process maintenance programs.

The results of evaluating the shutdown maintenance activities against the three main critical factors will be a set of consequences, which represent the third element of the SMWAM model. Each one of these consequences could be represented by one of the following:

■ The evaluated shutdown maintenance activity is a safety critical activity, and it should not be transferred to other in-process maintenance programs since it will cause harm or danger to the outside environment or to the workplace when they are performed by other in-process maintenance programs.

■ The evaluated shutdown maintenance activity is an operationally critical activity, and it should not be transferred to other in-process maintenance programs since it will cause partial or full unplanned shutdowns or reduce the quality of the final product when they are performed by other in-process maintenance programs.

■ The evaluated shutdown maintenance activity is a costly critical activity, and it should not be transferred to other in-process maintenance programs since the cost of performing this activity by other in-process maintenance programs is greater than performing it in the shutdown maintenance that will cause an increase in the overall maintenance cost.

■ The evaluated shutdown maintenance activity is a noncritical activity, and it should be transferred to other in-process maintenance programs since it will not cause any safety or operational problems and will not increase the overall maintenance cost.

10.5 Structure of SMWAM

The best way to describe the structure of this model is through its main steps, which are illustrated in the logical decision tree in Figure 10.1.

10.5.1 Step 1: Evaluating the Inclusion of Maintenance Activities on the Shutdown Maintenance List According to Operational Issues

This step of the model starts after receiving the work request forms that are related to shutdown maintenance from each unit and department and after the completion of the initial candidate list of the shutdown maintenance activities. It starts first by making each activity in this list answer to the questions that are related to the operational issues in order to evaluate its inclusion in the list of shutdown maintenance activities. In general, in this step two main questions are asked:

■ Does this activity cause full shutdown to the plant, if it is performed by another in-process maintenance program?
■ Does this activity cause partial shutdown to the plant, if it is performed by another in-process maintenance program?

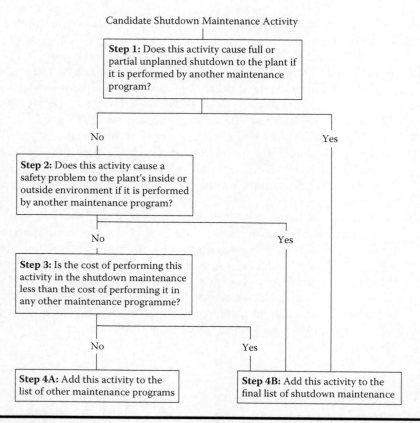

Candidate Shutdown Maintenance Activity

Step 1: Does this activity cause full or partial unplanned shutdown to the plant if it is performed by another maintenance program?

No   Yes

Step 2: Does this activity cause a safety problem to the plant's inside or outside environment if it is performed by another maintenance program?

No   Yes

Step 3: Is the cost of performing this activity in the shutdown maintenance less than the cost of performing it in any other maintenance programme?

No   Yes

Step 4A: Add this activity to the list of other maintenance programs

Step 4B: Add this activity to the final list of shutdown maintenance

Figure 10.1 The structure of the shutdown maintenance scope of work assessment model (SMWAM).

If the answer to any one of these questions is yes, then the evaluated activity will be included in the final list of the shutdown maintenance activities. Otherwise, it will be moved to the next step for further evaluation.

These two general questions were used to represent the important issues about the plant operations; however, there could be more than these two issues, depending on how the plant operation is defined. In addition, in the second question of this evaluation, there is no specific percentage figure that illustrates the concerned level of the partial shutdown maintenance. This also depends on how the word *partial* is being defined by the people in the plant where the model is to be applied.

The way to perform and facilitate the above evaluation is by using a special table that summarizes the results of the evaluation and simplifies the process of implementation. In addition, this table serves as a documentation tool for the evaluation results. A sample of this table is illustrated in Table 10.1.

Table 10.1   The Evaluation Table for the First Step in the Structure of the SMWAM Model

Shutdown Maintenance Scope of Work Assessment Model (SMWAM)

Step 1: Evaluating the inclusion of maintenance activities into the shutdown maintenance list according to the operational issues.	**Date:**

Plant:	**Plant ID:**

Analysts:

List of Issues

1. If the activity is performed by another maintenance program, does it cause full shutdown maintenance to the plant?
2. If the activity is performed by another maintenance program, does it cause 20% partial shutdown to the plant?

Activity	Answers		The Result
	Question #1	Question #2	
201-V1/Change catalyst	N	Y	SH
201-V1/Inspection of NHT	N	Y	SH
201-E1/Inspection of the combine feed–effluent exchanger	N	Y	SH
201-E1/Carry out DDT	N	Y	SH

(Continued)

Table 10.1 (*Continued*) The Evaluation Table for the First step in the Structure of the SMWAM Model

Activity	Answers		The Result
	Question #1	Question #2	
201-H1/Test and maintain charge heater	N	N	NP
201-V2/Inspection of kero-reactor	N	N	NP
301-C1.CT1/PM Activity for recycling gas compressor	N	Y	SH
Instrumentation and electrical jobs Class A	N	Y	SH
Overhaul safety valves Class A	N	Y	SH
Instrumentation and electrical jobs Class B1	N	N	NP
Overhaul safety valves Class B	N	N	NP
Repair/replace vents and drains Class B2	N	N	NP
301/Renewal of corrosion probes	N	N	NP
301/Replacement of pipes for 301 heater	N	N	NP
411/Replacement of pipes for 411 heater	N	Y	SH
Tie-in piping for new project	N	Y	SH

In this table, there is a space available in the center for writing questions related to the plant operation issues. In the general case mentioned earlier, the two questions mentioned have filled this space. Below this space, there are three main columns. The first column on the left-hand side is for listing the activities. Following this column is the column that is dedicated to the answers to the questions. This column is further subdivided into several columns according to the number of questions. Each one of these columns will be dedicated to register the answer to each question, which means that the number of these columns should match the number of questions. The last column in this table represents the results column where the results of analyzing the answers in the preceding subcolumns are registered.

The analysis process of the answers to the questions will be of two types. The first type is to analyze the answers logically by using the logic expressions such as (OR), (AND), or a combination of both. The second type is the one that analyses the answers arithmetically by using mathematical equations. The choice of the analysis process type depends on the type of data (quantitative/qualitative (yes/no)) that is related to each issue and the relationship between the issues. In general, the arithmetical analysis is used when all the issues are measured quantitatively and logic analysis is used when all issues are measured qualitatively (yes/no) or when some of these issues are measured quantitatively and the other some qualitatively. In the preceding general case, the two operational issues are mutually exclusive and the data related to them is qualitative in nature, which facilitate the use of the logic expressions for the analysis of the answers to the questions that are related to these issues. The logic expression used in this case is the logic expression (OR) since at least one of these issues needs to be satisfied in order to include the evaluated maintenance activity in the shutdown maintenance list. This means that if the answer to any one of the questions is (Yes), then the result in the last column (the results column) is (Yes) and that evaluated maintenance activity should be included in the shutdown maintenance list of activities. Otherwise, the result in the last column is (No) and that particular maintenance activity should move to the next step for further evaluation.

Note: SH means the activity is a shutdown maintenance activity. NP means the activity did not pass the test.

In addition to the aforementioned spaces, the table has several other spaces dedicated to information such as the analyst (analysts) name, the plant name and ID, the date of the evaluation, and the evaluation step title.

10.5.2 Step 2: Evaluating Inclusion of Remaining Maintenance Activities from First Step into Shutdown Maintenance List According to Safety Issues

In this step, the remaining maintenance activities from the first step that are not included in the final list of the shutdown maintenance activities are evaluated according to the safety issues. This evaluation is carried out by examining each one of these activities against questions related to safety issues. In general, the examination of these activities is against the following questions:

- Does this activity cause any death or injury to the employees in the plant, if it is performed by another in-process maintenance program?
- Does this activity cause any death or injury or harm to the people outside the plant, if it is performed by another in-process maintenance program?
- Does this activity cause any danger or damage to the physical workplace in the plant, if it is performed by another in-process maintenance program?
- Does this activity cause any harm to the outside environment, if it is performed by another in-process maintenance program?

If the answer to any one of these questions is (Yes), then the evaluated activity will be included in the final list of the shutdown maintenance activities. Otherwise, it will be moved to the next step for further evaluation. These four questions were used to represent the general safety issues; however, there could be more than these issues depending on how the plant safety is defined.

Similar to the first step, an analysis table and an analysis process are used in this step in order to facilitate the preceding evaluation and to summarize the results. In the foregoing general case of this step, the table used includes four questions related to safety issues and the answers to these questions are analyzed logically by using the logic expressions (OR) since at least one of these issues needs to be satisfied in order to include the evaluated maintenance activity in the shutdown maintenance list. This means that if the answer to any one of these questions is (Yes), then the result of the analysis that will be recorded in the last column is (Yes), which in turns means that the evaluated maintenance activity should be included in the shutdown maintenance list of the activities. Otherwise, the result of the analysis that will be recorded in the last column is (No) and that particular maintenance activity should move to the next step for further evaluation. A sample of this table is illustrated in Table 10.2.

10.5.3 Step 3: Evaluating Inclusion of Remaining Maintenance Activities from the Second Step into Shutdown Maintenance List According to Economic Issues

In this step, the maintenance activities that remain from the last step and which are not included in the final list of the shutdown maintenance activities are evaluated according to economic issues. This evaluation is carried out by comparing the total cost of performing each activity in the list when it is a part of the shutdown maintenance program with the total cost of performing it in any other maintenance program. In both cases, the total cost includes the direct maintenance cost (direct materials, spare parts, labor, tools, and equipment costs), the indirect maintenance

Table 10.2   The Evaluation Table for the Second Step in the Structure of the SMWAM Model

Shutdown Maintenance Scope of Work Assessment Model (SMWAM)

Step 2: Evaluating the inclusion of maintenance activities into the shutdown maintenance list according to the safety issues.	Date:

Plant:	Plant ID:

Analysis:

List of Issues

1. If this activity is performed by another in-process maintenance program, does it cause any death or injury to the employees in the plant?

2. If this activity is performed by another in-process maintenance program, does it cause any death or injury or harm to the people outside the plant?

3. If this activity is performed by another in-process maintenance program, does it cause any danger or damage to the physical workplace in the plant?

4. If this activity is performed by another in-process maintenance program, does it cause any harm to the outside environment?

Activity	Answers				The Result
	Question #1	Question #2	Question #3	Question #4	
210-H1/Test and maintain charge heater	N	N	N	N	NP
210-V2/Charge catalyst	N	N	N	N	NP

(Continued)

Table 10.2 (*Continued*) The Evaluation Table for the Second Step in the Structure of the SMWAM Model

Activity	Answers				The Result
	Question #1	Question #2	Question #3	Question #4	
210-V2/Inspection of kero-reactor	N	N	N	N	NP
Instrumentation and electrical jobs Class B1	N	N	N	N	NP
Instrumentation and electrical jobs Class B2	N	N	Y	N	SH
Overhaul safety valves Class B	N	N	N	N	NP
Repair/replace vents and drains Class B1	N	N	N	N	NP
Repair/replace vents and drains Class B2	Y	Y	Y	Y	SH
310/Renewal of corrosion probes	N	N	N	N	NP
310/Replacement of pipes for 301 heater	N	N	N	N	NP
Tag jobs Class B1	N	N	N	N	NP
Tag jobs Class B2	Y	Y	Y	Y	SH

cost (indirect materials, spare parts, labor, tools, and equipment costs), the overhead cost (other expenses used for carrying out maintenance activities), and the losses (the loss in production as a result of carrying maintenance activities, loss of sales, loss of quality). If the result of this comparison shows that the total cost of performing a particular activity in the shutdown maintenance program is less than the total cost of performing the same activity in any other maintenance program, then the evaluated activity will be included in the final list of the shutdown maintenance activities. Otherwise, it will be moved to the list of the maintenance program where the total cost of performing this activity is less than the total cost of performing it in the shutdown maintenance.

The way to simplify the use of the above evaluation is through the use of a similar table to the one used in the first and the second steps. However, each question in this table will be related to the total cost of performing a particular activity in each possible maintenance program, and one of these questions is related to the total cost of performing the same activity in the shutdown maintenance. In addition, the evaluation in this table is based on finding the maximum saving (Y_i) in the total cost of performing a maintenance activity (i). The way to achieve (Y_i) is through searching for the maximum difference between the total cost of performing a maintenance activity (i) in the shutdown maintenance and the total cost of performing the same activity (k) in each possible maintenance program (j). The mathematical equation that is used for the calculation of (Y_i) is the following:

$$Y_i = \max\nolimits_{all\ j} (X_i - Z_{kj}) \quad i = 1,2,\dots,m; \quad j = 1,2,\dots,n \quad \text{and} \quad k = 1,2,\dots,pj \quad (10.1)$$

where

n = The total number of maintenance programs.

m = The total number of remaining maintenance activities from the first and second steps.

pj = The total number of maintenance activities in the j maintenance program.

Y_i = The maximum savings in the cost of performing the i activity.

X_i = The total cost of performing i activity in the shutdown maintenance program.

Z_{kj} = The total cost of performing k activity in the j maintenance program.

The total cost (X_i) in Equation 10.1 includes the maintenance cost (C_i), the losses (L_i) that are caused as a result of performing activity (i) in the shutdown maintenance, and the other cost elements (OC_i) that are associated with performing activity (i) in the shutdown maintenance. In other words, the total cost of performing activity (i) in the shutdown maintenance is equal to:

$$X_i = C_i + L_i + OC_i \quad (10.2)$$

The maintenance cost (C_i) in this equation is equal to the summation of the direct maintenance cost (DC_i), indirect maintenance cost (IDC_i), and the overhead cost (OVC_i) that are associated with performing the i_{th} activity in the shutdown maintenance. In particular, this maintenance cost (C_i) is equal to:

$$C_i = DC_i + IDC_i + OVC_i \qquad (10.3)$$

In addition, the losses (L_i) in Equation 10.2 represent different types of losses such as the loss of production (LP_i), the loss of sales (LS_i), and the loss of quality (LQ_i) that could happen as a result of performing activity (i) in the shutdown maintenance and it is expressed as follows:

$$L_i = LP_i + LS_i + LQ_i \qquad (10.4)$$

In addition to the total cost (X_i) in Equation 10.1, the total cost (Z_{kj}) also includes the maintenance cost (C_{kj}), the losses (L_{kj}) that are caused as a result of performing the k_{th} activity in the j_{th} maintenance program, and the other cost elements (OC_{kj}) that are associated with performing maintenance activity (k) in maintenance program (j). In other words, the total cost of performing activity (k) in maintenance program (j) is equal to:

$$Z_{kj} = C_{kj} + L_{kj} + OC_{kj} \qquad (10.5)$$

The maintenance cost (C_{kj}) in this equation is equal to the summation of the direct maintenance cost (DC_{kj}), the indirect maintenance cost (IDC_{kj}), and the overhead cost (OVC_{kj}) that are associated with performing the k activity in the j maintenance program. In particular, this maintenance cost is equal to:

$$C_{kj} = DC_{kj} + IDC_{kj} + OVC_{kj} \qquad (10.6)$$

In addition, the losses (L_{kj}) in Equation 10.5 represent deferent types of losses such as the loss of production (LP_{kj}), the loss of sales (LS_{kj}), and the loss of quality (LQ_{kj}) that could happen as a result of performing the k activity in the j maintenance program and it is expressed as follows:

$$L_{kj} = LP_{kj} + LS_{kj} + LQ_{kj} \qquad (10.7)$$

The result of this process (Y_i) will be recorded in the last column of the table, which is subdivided into two columns. One of them is dedicated to the result (Y_i), and the other is dedicated for recording the optimum maintenance program for performing the maintenance activity. The result (Y_i) and its interpretation could be one of the following:

■ The result (Y_i) is negative figure, which means that the particular maintenance activity which this figure is related to should be included in the shutdown maintenance list of activities.

- The result (Y_i) is zero, which means that the particular maintenance activity which this figure is related to could be included in either the shutdown maintenance list or in the list of the in-process maintenance program in which this activity has maintenance cost and loss of production that are similar to their counterpart of the shutdown maintenance.
- The result (Y_i) is positive, which means that the particular maintenance activity which this figure is belong to should be included in the list of the in-process maintenance program which has the lowest total maintenance cost and loss of production. A sample of this table is illustrated by Table 10.3.

Note: The cost figures in the middle of table have been erased because they were confidential. OP means other maintenance program.

10.5.4 Step 4: Preparing Final List of Shutdown Maintenance Activities

After the completion of the whole evaluation process by performing the last step, two lists of maintenance activities will be available. One of these lists is the final list of shutdown maintenance activities, which includes those activities that were confirmed to be shutdown maintenance activities and which successfully have been passed the evaluation process. In addition, the list also includes the reasons behind the selection of these maintenance activities. These types of information are represented in this list in a form of a table that has two main columns. The first one of these columns is dedicated for recording those maintenance activities that were confirmed to be shutdown maintenance activities. The second column is subdivided into three columns, each of which is dedicated for representing one of the following reasons for the inclusion of a particular activity in the list.

- The maintenance activity at hand is an operationally critical activity.
- The maintenance activity at hand is a safety critical activity.
- The maintenance activity at hands is an economically critical activity.

The other list that will be available after the completion of the preceding evaluation process is the list of maintenance activities that are related to the other in-process maintenance programs. This list includes those activities that failed to pass the evaluation process, and that are not included in the list of shutdown maintenance activities. It also includes the suggested in-process maintenance programs in which these activities will be performed.

The way in which these types of information are represented is by using a table that consists of two columns. The first column is dedicated for recording each maintenance activity in the preceding list of activities, and the second column is dedicated for recording the type of corresponding in-process maintenance program to that particular activity. These two lists are illustrated in Tables 10.4 and 10.5, respectively.

Table 10.3 The Evaluation Table for the Third Step in the Structure of the Model

Shutdown Maintenance Scope of Work Assessment Model (SMWAM)				
Step 3: Evaluating the inclusion of the remaining maintenance activities from the second step into the shutdown maintenance list according to the economic issues	**Date:**			
Plant:	**Plant ID:**			
Analysis:				
List of Issues 1. What is the total cost of performance the activity in the shutdown maintenance program? 2. What is the total cost of performance the activity in the *first alternative* maintenance program?				
	Answers			**The Type of Maintenance Program**
Activity	**Q #1**	**Q #2**	**The Result**	
210 H1/Test and maintain charge heater			304400	OP
210-V2/Change Catalyst			327293	OP
210-V2/Inspection of kero-reactor			307901	OP
Instrumentation and electrical jobs Class B1			106485	OP
Overhaul safety values Class B			−15000	SH
Tag jobs Class B1			88365	OP
Repair/replace vents and drains Class B1			205329	OP
301/Renewal of corrosion probes			150530	OP
301/Replacement of pipes for 301 heater			136600	OP

Table 10.4   The Final List of Shutdown Maintenance Activities

Shutdown Maintenance Scope of Work Assessment Model (SMWAM)

Step 4: Preparing the final list of shutdown maintenance activities:

Date:

Plant:

Plant ID:

Analysis:

Activity	Operationally Critical	Safety Critical	Economically Critical
210-V2/Change catalyst	Y		
210-V2/Inspection of the NHT	Y		
201-E1/Inspection of the combine feed–effluent exchanger	Y		
201-H1/Inspection of naphtha hydro-treater	Y		
201-E1/Carry out DDT	Y		
301-V1/Change catalyst reactor #1	Y		
301-V2/Change catalyst reactor #2	Y		
301-V3/Change catalyst reactor #3	Y		
301-H1,2,3/Inspection of plat charge	Y		

(Continued)

Table 10.4 (Continued)   The Final List of Shutdown Maintenance Activities

Activity	Operationally Critical	Safety Critical	Economically Critical
301-H1,2,3/Replacement of dampers	Y		
301-C1.CT1/PM Activity for recycle gas compressor	Y		
411-H1/Repair floor	Y		
Instrumentation and electrical jobs Class A	Y		
Instrumentation and electrical jobs Class B2	Y		
Overhaul safety valves Class A		Y	
Overhaul safety valves Class B			Y
Repair/replace vents and drains Class A	Y		
Repair/replace vents and drains Class B2		Y	
Tag jobs Class A	Y		
Tag jobs Class B		Y	
411/Replacement of pipes for 411 heater	Y		
Tie-in piping for new project	Y		

Table 10.5   The List of Other Maintenance Program Activities

Shutdown Maintenance Scope of Work Assessment Model (SMWAM)	
Step 4: Preparing the final list of shutdown maintenance activities	**Date:**
Plant:	**Plant ID:**
Analysis:	
Activity	**Recommended Maintenance Program**
210 H1/Test and maintain charge heater	Preventive maintenance
210-V2/Change catalyst	Preventive maintenance
210-V2/Inspection of kero-reactor	Preventive maintenance
Instrumentation and electrical jobs Class B1	Preventive maintenance
Tag jobs Class B1	Preventive maintenance
Repair/replace vents and drains Class B1	Preventive maintenance
301/Renewal of corrosion probes	Preventive maintenance
301/Replacement pipes for 301 heater	Preventive maintenance

10.6 How to Use SMWAM

The optimum way to use the model starts with forming a shutdown maintenance scope of the work assessment team. In this team, a representative from each discipline that is related to the plant operational, safety, and maintenance economic issues such as the plant maintenance personnel, operations personnel, safety personnel, inspection personnel, technical and service personnel, shutdown maintenance planner, and any other related personnel is included. The reason behind this is to gain maximum benefit from their experience in determining the plant operational, safety, and maintenance economic issues (criteria) and the limits that will be used in the model evaluation process. In addition, the involvement of such people is necessary and recommended in order to gain their buy-in to the model.

In addition to the formation of the team, the members of the team are provided with several sources of information that can help them to identify the different operational, safety, and maintenance economic issues (criteria) and their limits. Some of these sources of information are the following:

- Support financial data records and files that will provide information on the direct, indirect, and overhead costs of the maintenance activities which in turn represents valuable details in the determination of the economic issues and their limits.

■ Shutdown maintenance records and files that will provide information on the time required to perform each maintenance activity which in turn helps in calculating the loss in production that is caused by performing each activity. In addition, these records and files help in the determination of the nonacceptable level of the partial shutdown maintenance and also in the determination of the economic issues and their limits.

■ Safety records, files, and governmental and international safety regulations that will provide information on safety precautions needed to perform each maintenance activity and the level of the safety risk that is associated with failure to perform certain maintenance activities. This type of information will help in the determination of the safety issues and their limits.

■ System operation manuals, which provide information on how maintenance is related to the system, describe how systems are intended to function, how they relate to other systems, and what operational limits are employed. This information will help in the determination of the operational issues and their limits.

■ Any other source of information such as piping and instrumentation diagrams, process flow diagrams, work order systems, and equipment data sheet which the team identifies as important in the determination of the plant operational, safety, and maintenance economic issues.

After performing the SMWAM, with the final list of shutdown maintenance activities, the final list of those activities that will be performed by other in-process maintenance programs will be reviewed and approved by the management. At that time, both approved lists are ready to be executed.

10.7 Case Study

The preceding model has been applied to one of the oil refineries in Saudi Arabia, and the result of the application was the following:

■ A team was formed in order to perform the steps of the model. This team has representatives from the maintenance department, operations department, test and inspection department, and technical and service department, in addition to the shutdown maintenance planner and filed operation engineer.

■ Individual meetings were held with each group of team members who have the experience and the necessary information about the plant's operation, safety, or about the economic issues of the shutdown maintenance. The aim of these meetings was to determine the refinery's operational, safety, and maintenance economic issues. These issues have been represented by questions that have been written in the middle of Tables 10.1, 10.2, and 10.3, respectively.

- Several meetings were held with all team members in order to test each activity in the initial shutdown maintenance list against the preceding issues. The aim of the test was to identify those activities that are considered to be shutdown maintenance activities and those that are not. The result of this test on area (E) of the refinery, for example, was the following:

- Forty maintenance activities related to area (E) in the initial list of shutdown maintenance were tested against two operational issues as illustrated by Table 10.1. Twenty-eight of these activities passed the test and were classified as shutdown maintenance activities. The other twelve did not pass the test and were moved to the next step for further testing.

- The twelve activities, which did not pass the first step, were tested against four safety issues. The result was that three of these activities passed the test in this step and were classified as shutdown maintenance activities. The other did not pass the test, and they were moved to the next step for further testing. This test is illustrated by Table 10.2.

- The remaining nine activities from the last step were tested against two maintenance economic issues as illustrated by Table 10.3. The result was that eight of these activities did not pass the test, and they were classified as non-shutdown maintenance activities. They were transferred from the shutdown maintenance list to other in-process maintenance programs lists. The conse-quence of this action is a reduction of the shutdown maintenance duration of this area by approximately one-and-a-half day, which represents a saving in the total maintenance cost and the loss or production by approximately $1.6M.

- The final list of shutdown maintenance activities and another list of the activ-ities to be performed by the other in-process maintenance programs were formed and prepared by the team as a result of the preceding test. These lists of area (E) are illustrated by Tables 10.4 and 10.5, respectively.

10.8 Comparison between Conventional Way of Identifying Scope of Work of Shutdown Maintenance and SMWAM

The SMWAM and the present method of assessing the work of shutdown maintenance are similar in their aim since both of them are designed to achieve the proper assessment of the shutdown maintenance scope of work. However, there are several differences between these two approaches. These differences are summarized in the Table 10.6.

Table 10.6   The Comparison between the Conventional Way of Identifying the Scope of Work of Shutdown Maintenance and the Shutdown Maintenance Scope of Work Assessment Model (SMWAM)

Features	SMWAM	Conventional Way
Structure	The model has well-defined structure, which is represented by four steps.	There is no specific structure. The shutdown maintenance activities are specified through the collection of work request forms from production or service departments.
Flexibility	The model allows justifiable changes in the number of critical issues and the limits of these issues.	There is total flexibility, which is not restricted by any justification.
Documentation	The model is well documented through the use of tables that include all necessary information and justification regarding inclusion of certain maintenance activity in the shutdown maintenance list.	The documentation is weak. Sometimes it is represented by only a final list of shutdown maintenance activities.
Bases of selection process	The selection process is based on the issues related to the plant operation, safety, and maintenance cost.	The base of the selection process is changing, which sometimes causes some noncritical maintenance activities to be included in the shutdown maintenance list, and at other times it causes some critical maintenance activities to be excluded from the shutdown maintenance list.

10.9 Conclusion

The SMWAM that has been discussed earlier have several specific characteristics that make the model useful for reassessing the scope of work of shutdown maintenance in order to reduce the shutdown maintenance cost and the loss in production that is caused by shutdown maintenance. These characteristics are summarized as follows:

- The model is relatively complete and simple to use. It is complete in the sense that it includes important factors that affect the assessment process as identified by industrial maintenance managers. In addition, it is simple in the sense that it represents these factors in an understandable way.
- The preceding model is also easy to control. The user will have control of the issues that represent each factor affecting the selection process of the maintenance activities. In addition, this control can be exceeded to include specific limits for some issues such as the percentage that illustrates the concerned level of the partial shutdown maintenance.
- The model is an adaptive and flexible model. It is so in the sense that it has two forms of analysis to choose from that are related to the factors at hand. These are the logical and the arithmetical forms. In addition, an increase in the number of the factors that affect the selection process under this model is possible.
- The model is easy to communicate. The data required by the model use the same terms that managers are familiar with. In addition, the output from the model has been represented in a tabulated form, and lists that make important information such as the date of evaluation, the analyst (analysts) name, the results of the evaluation, and the evaluation steps to be easily followed and communicated by the related personnel.

Author

Dr. Adel Al-Shayea is a consultant industrial engineer (CE-SCE). He is a member of the Saudi Council of Engineers (SCE) Saudi Arabia as well as a member of several committees such as the national committee for the codification and standardization of operation and maintenance works.

He is currently the chairman of the Industrial Engineering Department at the College of Engineering, King Saud University. Previously, Dr. Al-Shayea was a consultant at King Saud University Rector's Office and also worked for SABIC Marketing Ltd. in Riyadh and at the Institute of Public Administration (IPA).

In addition, Dr. Al-Shayea is a consultant in King Abdullah Institute of Research and Consulting Studies. He participated and conducted several consultative works for governmental and private organizations. He also refereed several engineering works in Saudi Arabia.

References

Al-Najjar, B., and I. Alsyouf. 2003. Selecting the most efficient maintenance approach using fuzzy multiple criteria decision making. *International Journal of Production Economics* 84(1): 85–95.

Al-Shayea, A. 2003. Assessment of the Scope of Shutdown Maintenance Work. Ph.D. diss., Loughborough University.

Ashayeri, A., A. Teelen, and W. Selen. 1996. A production and maintenance planning model for the process industry. *International Journal of Production Research* 34(12): 3311–26.

Austin, L., and P. Ghandforoush. 1993. *Management Science for Decision Makers*. Saint Paul: West Publication Company.

Baughman, H., and A. Shah. 1996. Template planning for refinery shutdowns. *Proceedings of International Industrial Engineering Conference* 292–97.

Chareonsuk, C., N. Nagarur, and M. Tabucanon. 1997. A multicriteria approach to the selection of preventive maintenance intervals. *International Journal of Production Economics* 49: 55–64.

Dekker, R., and P. Scarf. 1998. On the impact of optimization models in maintenance decision making: The state of the art. *Reliability Engineering and Systems Safety* 60(2): 111–19.

Ghosh, D., and S. Roy. 2009. Maintenance optimization using probabilistic cost-benefit analysis. *Journal of Loss Prevention in the Process Industries.* 4(22): 403–7.

Goicoechea, A., D. Hasen, and L. Duckstein. 1982. *Multiobjective Decision Analysis with Engineering and Business Application*. New York: John Wiley & Sons.

Gupta, S., and J. Paisie. 1997. Reduce turnaround costs. *Hydrocarbon Processing* January: 67–74.

Kelly, A. 2006. *Plant Maintenance Management Set*. Oxford: Butterworth Heinemann.

Koppelman, J., and R. Faris. 1994. Optimize plant shutdowns. *Chemical Engineering Progress* May: 68–73.

Mathew, J., and C. Rajendran. 1993. Scheduling of maintenance activities in a sugar industry using simulation. *Computers in Industries* 21: 331–34.

Shuffelton, D. 1998. Outage management the way ahead? *The Nuclear Engineer* 39(3): 88–92.

Smith, R., and B. Hawkins. 2004. *Lean Maintenance*. Oxford: Butterworth Heinemann.

Chapter 11

Machine Failure Time Detection through Product Defects

Hazem J. Smadi

*Industrial Engineering Department, Jordan University
of Science and Technology, Irbid, Jordan*

Contents

11.1 Introduction

Throughout the years, the need for maintenance management has increased in order to efficiently manage and control maintenance activities vital to keep production running competitively. Maintenance cost is the second-largest cost in the operational budget of refineries; hence, proper management is required to reduce costs (Garag and Deshmukh 2006, Simeu-Abazi and Bouredji 2006).

Maintenance is defined as the combination of activities and actions that are required to control and supervise a system or a component to perform the intended functions. The main objective is to restore the system to a state in which it can perform the intended function. Reduction of breakdowns can be achieved through preventive maintenance, which is the activities that are performed to keep the system running before failures can occur (Simeu-Abazi and Bouredji 2006).

Maintenance policy addresses performance of maintenance activities. It is dependent on available resources and the level at which activities are done, based on the complexity of the system (systems level, subsystem, component, etc.). Maintenance policies are classified into three categories; corrective maintenance, preventive maintenance, and mixed (corrective and preventive) (Simeu-Abazi and Bouredji 2006).

11.2 Maintenance Policies

Corrective maintenance is called for when there is a breakdown or machine failure. It is unplanned and aimed at restoring a machine to its functional state in order to resume production. The maintenance activities that fix failures only are known as *breakdown maintenance*; *corrective maintenance* involves other possible failures anticipated when the breakdown occurs (Ahuja and Khamba 2008, Simeu-Abazi and Bouredji 2006).

Preventive maintenance is planned maintenance (with a time-line plan) that is performed based on a schedule. The objective is to reduce the probability of sudden breakdowns by keeping a good operational status for machines. This increases the availability and the reliability of machines. Preventive maintenance does not omit the need for corrective maintenance, although required maintenance should reduce the probability of the breakdowns and maintain production operational (Ahuja and Khamba 2008, Simeu-Abazi and Bouredji 2006).

Mixed maintenance policy is composed of both corrective and preventive policies at the same time. If a failure occurs before the scheduled day of preventive maintenance, corrective maintenance takes place (Ahuja and Khamba 2008, Simeu-Abazi and Bouredji 2006).

11.3 Quality and Maintenance

Companies recognize the importance of quality as key to competition and survival. In some organizations, quality has become a business strategy for success and growth. One of the main functions in an organization is production, which has been addressed for many years in research. Maintenance is also another vital function for maintaining quality, and this has not been the subject of many researches (Mehdi et al. 2010, Panagiotidou and Nenes 2009).

Through studying the literature, we see some new researches that addressed the problem of quality-maintenance integration. Most of the work is related to quality control or total-quality management tools. There is no direct link between a product's physical quality and the maintenance system. This is the scope of our discussion here.

This chapter focuses on building a preventive maintenance policy that considers actual product defects that are caused due to machine malfunction in order to predict the next failure. This is one way of establishing and developing a preventive maintenance policy that reduces the activities related to corrective maintenance taking place due to breakdowns.

11.4 Previous Related Work

A joint quality control and preventive maintenance policy for a production system producing conforming and nonconforming units has been developed by Mehdi et al. (2010). A single machine system has been developed with a proportion of nonconforming units on each lot compared to a threshold value to decide the system's maintenance. A buffer stuck is built when a threshold reaches a certain value after a machine has gone down for maintenance. The objective is to determine the buffer size and threshold values that minimize the total expected cost. Panagiotidou and Nenes (2009) developed a model for the economic design of a variable parameter for a Shewhart control chart that monitors the process's mean. Beside the quality shift in the production, a failure may occur. Removing the quality shifts and improving in the product quality is considered preventive maintenance as it improves the reliability of the process and reduces the probability of failure. The scheme parameters minimize the expected quality and maintenance costs determined by the model. The result of the optimal cost is compared against the optimum cost of a fixed parameter chart.

Panagiotidou and Tagaras (2007) presented an economic model for the optimization of preventive maintenance in a production process with two quality states; a control state may shift to an out-of-control state before a failure occurs or a scheduled preventive maintenance takes place. Each state has a different failure rate and revenue. The time to perform preventive maintenance is identified by two critical values of equipment age that have been found optimal. Possible extensions may consider inaccurate process monitoring, imperfect maintenance actions, and gradual quality deterioration systems.

11.5 Proposed Methodology

In general, failures are not deterministic, they are stochastic; a failure can occur any time with respect to certain stochastic behavior. In general, repairs for these failures are not deterministic as well; they follow also certain stochastic behavior. Failure and repair data are essential for preventive maintenance policy development. The methodology flowchart is illustrated in Figure 11.1. The feedback link is to enhance and update some data that affect the preventive maintenance policy (Smadi 2011).

11.5.1 Stage 1: Data Preparation

Data are not available in the same required format and shape to be analyzed. As this methodology deals with statistical data, an important initial step is data preparation. It is very important to be careful at initial preparation stages before

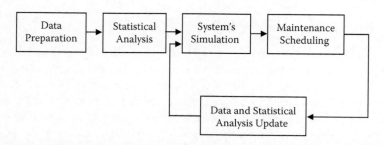

Figure 11.1   The methodology flowchart.

performing any analysis activity, as the final results are affected by the quality of the data. The data preparation is to be performed for two different type of data; data that are related to machines and data that are related to products.

11.5.1.1 Machine Data Preparation

The machine data are related to failure and repair behaviors of the machine. Historical maintenance orders and records may show the time that the machine fails (failure occurs), the type of failure that occurred, and the time the machine is back to work (repair time). It is beyond the scope of this chapter to study the elements of these records. The data are important to represent the behavior of each machine to be able to model a preventive policy.

11.5.1.2 Product Data Preparation

The preventive policy to be developed will consider the integration between maintenance and product quality; hence, some data that are related to product's quality are required. Usually, nonconformity is reported when there is a defect in the production. The inspector and the production team can identify the reason for the occurred defect; it can be due to materials, operations parameters, the operator, and machine malfunction. This step is interested in data (time) for product defects that are due to machine malfunction.

11.5.2 Stage 2: Statistical Analysis

Each type of data in stage 1 is analyzed in this stage. The statistical analysis is in two forms; distribution fitting and multiple variable regression analysis.

11.5.2.1 Distribution Fitting

In general, failure time is fitted using time between failures, so mean time between failures is the mean of the fitted distribution. The repair time is fitted as the period of time to perform a repair job. Statistical distributions can be parametric or nonparametric. It is preferred to work with parametric distribution as they are easier and less complex. In some cases, it is not possible to fit the data into a parametric distribution; in this case, nonparametric distribution is to be considered.

11.5.2.2 Multiple Variable Regression Analysis

Multiple variable regression analysis is to be conducted to predict the next coming failure, given that a product defect occurred. This entails the integration between maintenance and product quality; this link has not been addressed before. Multiple variable regression ends up with a long-term relationship between independent variables and a dependent variable; hence, a future value for the dependent variable can be predicted using certain values of the independent variables and this relationship. The term *multiple variable* is related to more than one variable as independent variables.

The independent variables in the proposed regression model are the time when a product's defect occurs and the type of failures that are associated to each machine. The dependent variable is the time of the next failure. The model assumes that there are significant effects of the defect time and the type of the predicted failure. After conducted regression analysis, the variables that are not significant will be eliminated from the model. There is one regression model for each machine. This analysis is separated from the analysis in step 1, but the result of step 1 and step 2 will work together in stage 3 of the proposed methodology.

11.5.3 Stage 3: System Simulation

Simulation is an effective tool for modification experiments. In these, a simulation model that matches the current system is developed. Later, some modifications are introduced in this model for experimentation purposes. A simulation model that addresses the current system is developed and is called an "as is" model. This model is considered as a base model that is used later for modification experimentation. Before proposing any modifications, this simulation model should be verified and validated.

Modifications are introduced into the "as is" simulation model and are tested for maintenance planning and scheduling. A new simulation model that is called a "defect time" simulation model is developed. The historical defect times DTs and regression models for each machine are introduced in the "as is" simulation model to develop the "defect time" simulation model. This model is used to generate DTs using historical data. These generated DTs are used in the same simulation model to calculate predicted failure times PFTs using regression models that have been developed previously. The maintenance schedule is developed according to these PFTs.

11.5.4 Stage 4: Maintenance Scheduling

Maintenance scheduling has been extensively reviewed in maintenance management literature. The objective is to merge production schedules with maintenance schedules, which will increase the overall efficiency of systems. Any scheduling methodology can be incorporated. The proposed methodology entails a dynamic preventive maintenance plan as it predicts future failures as time is running, based on the quality status. Generated PFTs form "defect time" simulation models used for maintenance scheduling. The last stage in the methodology addresses updating of the current probability distribution and regression model to better represent the system.

11.5.5 Stage 5: Data and Statistical Analysis Update

As the system is operating, failures occur and product defects are detected, changing the current data of failure time and product defect time, which may result in a different probability distribution and different regression model from what is currently available. To accurately represent the current status of the system, data should be updated and statistical analysis should be performed continuously. This continuous updating is a feedback process that will update the maintenance plan with respect to the behavior of the system.

11.6 Case Study

The methodology has been developed as a generic platform for preventive maintenance planning that can be implemented in vast range of industries. The required data in the methodology can be obtained in most organization. As a case study, the methodology was implemented to show the results for a real production line maintenance problem. The case study is a local company that has several plants across the United States. A production line that manufactures the majority of the company's products has been chosen for this analysis. The production line consists of six machines in a series, performing metal cutting operations. Figure 11.2 illustrates the process flow and the processing time of each operation.

Machine failures are classified into four categories: mechanical, electrical, hydraulic, and coolant. Table 11.1 lists the type of failure for each machine in the production line. Once a failure occurs, a maintenance worker records the time (down-time) of the failure and classifies the type of the failure according to the previous four categories. When maintenance is completed, the maintenance worker records the time (up-time), hence, the failure time and the time to repair are recorded. When

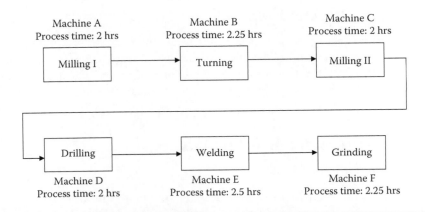

Figure 11.2 Production line operations and processing times.

Table 11.1 Existence of the Type of Failures for Each Machine

Machine	Type of Failures (Yes: Exists; No: Does Not Exist)			
	Mechanical	Electrical	Hydraulic	Coolant
Machine A	Yes	Yes	Yes	Yes
Machine B	Yes	Yes	Yes	Yes
Machine C	Yes	Yes	Yes	Yes
Machine D	Yes	Yes	No	Yes
Machine E	Yes	Yes	No	No
Machine F	Yes	Yes	Yes	No

a quality control department finds a product defect through inspection, the defect is documented through a nonconformity record that shows the reason of the defect. In this study we are interested in the defects that are due to machine malfunction.

Currently, the maintenance that is performed is only a routine preventive maintenance that deals with predetermined components to be checked every two to four weeks, based on the recommendations of the manufacturer of the machine for oil change, calibration, and so on. It has been found that this is not an efficient preventive maintenance plan that reduces corrective maintenance dramatically. Following is an implementation of the proposed methodology to improve the preventive maintenance efficiency for the production line of the case study. For simplification the implementation is shown for machine A only. The implementation for other machines follows the same methodology.

11.7 Stage 1: Data Preparation

The number of data points that is available is not the same for each machine; it depends on the number of hours a machine has been operating, and the number of data points for each different type of failure for each machine is not the same, depending on the frequency of the type of failure that can be found.

11.8 Stage 2: Statistical Analysis

11.8.1 Distribution Fitting

Distribution fitting was conducted for failure and repair data for each type of failure and each machine. The available data for failure are the exact failure time, taking the difference between each record (time) and the previous record. Table 11.2 lists the

Table 11.2   The PDF and the CDF for the Distributions Used in Goodness of Fit Test

Distribution	Parameters		f(x) PDF	F(x) CDF
Weibull	α	B	$\dfrac{\alpha}{\beta}\left(\dfrac{x}{\beta}\right)^{\alpha-1}\exp\left(-\left(\dfrac{x}{\beta}\right)^{\alpha}\right)$	$1-\exp\left(-\left(\dfrac{x}{\beta}\right)^{\alpha}\right)$
	Shape parameter	Scale parameter		
Gamma	α	B	$\dfrac{x^{\alpha-1}}{\beta^{\alpha}\,\Gamma(\alpha)}\exp\left(\dfrac{-x}{\beta}\right)$	$\dfrac{\Gamma_{x/\beta}(\alpha)}{\Gamma(\alpha)}$
	Shape parameter	Scale parameter		

probability density function PDF and the cumulative distribution function CDF for the distributions used in the goodness of fit test. As an example, the following is the distribution fitting for failure and repair time for machine A for a mechanical failure.

Chi-Square Test (goodness-of-fit test)

H_o: Time between failures for a mechanical failure follows a Weibull distribution with $\alpha = 0.583$ and $\beta = 154.17$

H_1: Time between failures for a mechanical failure does not follow a Weibull distribution

Data statistics:

Minimum = 0.366, Maximum = 3042.823,
Mean = 241.193, Standard deviation = 439.217

Table 11.3 lists the observed and theoretical frequencies for the chi-square test of mechanical failures for machine A. The table shows columns for the adjusted frequencies after combining data frequency that is less than 5.

The degree of freedom is $4 - 2 - 1 = 1$, χ^2 statistic = 1.246, χ^2 critical = 3.841, and p-value = 0.264 for a significance level of $\alpha = 0.05$; the null hypothesis cannot be rejected, and the data follows a Weibull distribution with the given parameters. The histogram with the fitted distribution is illustrated in Figure 11.3.

A chi-square test has been used for a sample size of more than 60, and the Kolmogorov–Smirnov test has been used for smaller samples. Table 11.4 lists the result of the goodness-of-fit test for failure and repair times for each type of failure for machine A.

11.8.2 Multiple Variable Regression Analysis

Regression analysis has been used to predict the next failure after a product's defect is detected. The dependent variable to predict is Y, the time of failure, where the

Table 11.3 The Observed and Theoretical Frequencies for Chi-Square Test of Mechanical Failures for Machine A

Class	Lower Bound	Upper Bound	Frequency (Data)	Adjusted Frequency (Data)	Frequency (Distribution)	Adjusted Frequency (Distribution)	Chi-square
1	0.000	238.462	54	54	55.070	55.070	0.021
2	238.462	476.923	12	12	9.918	9.918	0.361
3	476.923	715.385	5	5	4.433	6.804	0.650
4	715.385	953.846	0		2.371		
5	953.846	1192.308	2	5	1.393	3.968	0.213
6	1192.308	1430.769	2		0.870		
7	1430.769	1669.231	0		0.567		
8	1669.231	1907.692	0		0.382		
9	1907.692	2146.154	0		0.264		
10	2146.154	2384.615	0		0.187		
11	2384.615	2623.077	0		0.134		
12	2623.077	2861.538	0		0.098		
13	2861.538	3100.000	1		0.072		
						Total	1.246

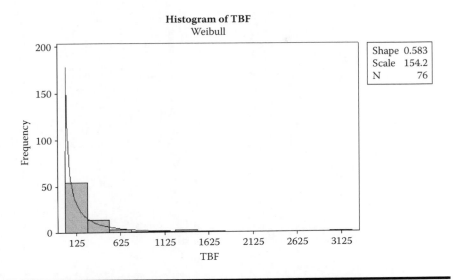

Figure 11.3 Histogram of TBF for mechanical failure for machine A with fitted distribution.

independent variables are the time of product's defect and the type of failure. The required data have been organized in a suitable format. For each time of product defect, there is more than one possible time of failure associated with each type of failure. In order to represent the type of failure in the regression model, dummy coding is incorporated. Dummy coding uses ones and zeros to code variables. The number of contrasts that are used is less than the number of variables by one. Variable one is given 1 in the first contrast, and all other variables are set to 0; in the second contrast variable two is set to 1 and all other variables are set to 0 and so forth (Davis 2010). Following is the regression analysis result for the six machines in the production line. Regression analysis was conducted using the statistical package Minitab 15.

The original data that are used for regression modeling of machine A is listed in Table 11.5. The failure types have been coded using dummy variables. The dummy coding according to the types of failures of machine A is listed in Table 11.6. Dummy variable 1 (DV1) receives a value of 1 when the failure is coolant; dummy variable 2 (DV2) receives a value of 1 when the failure is electrical; dummy variable 3 (DV3) receives a value of 1 when the failure is hydraulic, and when the failure is mechanical, the value of all dummy variables is 0.

R-Square value is considered a high value that represents a high confidence of the regression model. Table 11.7 lists the standard error and R-square values.

Table 11.8 lists the ANOVA table with F statistic equals 56.81, which is higher than the F table critical value (4.53) at $\alpha = 0.05$. The F test shows that the model is significant.

Table 11.4 The Result of the Goodness-of-Fit Test for Machine A

Failure			Type of Failure			
			Mechanical	Electrical	Hydraulic	Coolant
H	No. of Points		76	68	18	50
	H₀		Data follow a Weibull distribution	Data follow a Gamma distribution	Data follow a Weibull distribution	Data follows Gamma distribution
	Hₐ		Data do not follow a Weibull distribution	Data do not follow a Gamma distribution	Data do not follow a Weibull distribution	Data do not follow a Gamma distribution
Parameters			α 0.583	α 0.23	A 0.321	α 0.25
			β 154.17	β 1205.9	B 186.79	β 1622
GOF	Test		Chi-square	Chi-square	Kolmogorov–Smirnov	Kolmogorov–Smirnov
		Critical	3.841	9.488		
		Statistic	1.246	5.475	0.143	0.153
		p-value	0.264	0.242	0.826	0.174
Result			H₀ cannot be rejected	H₀ cannot be rejected	H₀ cannot be rejected	H₀ cannot be rejected

Repair			75	69	19	51
No. of points			75	69	19	51
H	H_o		Data follow a Gamma distribution	Data follow a Gamma distribution	Data follow a Gamma distribution	Data follow a Weibull distribution
	H_a		Data do not follow a Gamma distribution	Data do not follow a Gamma distribution	Data do not follow a Gamma distribution	Data do not follow a Weibull distribution
Parameters		α	0.672	0.697	α — 0.631	A — 1.109
		β	2.295	2.302	β — 2.079	B — 1.281
GOF	Test		Chi-square	Chi-square	Kolmogorov–Smirnov	Kolmogorov–Smirnov
	Critical		9.488	9.488		
	Statistic		5.2	4.005	0.268	0.171
	p-value		0.267	0.405	0.107	0.091
Result			H_o cannot be rejected	H_o cannot be rejected	H_o cannot be rejected	H_o cannot be rejected

Table 11.5   Data Used for Regression Modeling Machine A

Defect Time	Failure Time	Failure Type
0	3072	Coolant
0	3456	Electrical
0	2712	Mechanical
5688	6384	Coolant
5688	5880	Electrical
5688	8016	Hydraulic
5688	5880	Mechanical
14064	14904	Coolant
14064	15072	Electrical
14064	14712	Hydraulic
14064	14784	Mechanical

Table 11.6   The Dummy Coding According to the Types of Failures of Machine A

Failure	DV1	DV2	DV3
Coolant	1	0	0
Electrical	0	1	0
Hydraulic	0	0	1
Mechanical	0	0	0

Table 11.7   Standard Error and *r*-square Values for Regression Analysis Machine A

Standard error (σ)	1073.87
R-Square	97.4%
R- (adjusted)	95.7%

Table 11.8   The ANOVA Table of Regression Analysis of Machine A

Source	DF	SS	MS	F	P
Regression	4	262036274	65509069	56.81	0
Residual error	6	6919176	1153196		
Total	10	268955450			

Table 11.9   The Regression Analysis Result for Machine A

Predictor	Coefficient	Standard Error (SE)	t-Statistic	p-Value
Constant	2182.3	730.4	2.99	0.024
Defect time	0.85202	0.05864	14.53	0
DV1	328	876.8	0.37	0.721
DV2	344	876.8	0.39	0.708
DV3	767.2	999.1	0.77	0.472

$F = 56.81 > F_{(0.05,4,6)} = 4.53$; the model is significant.

Table 11.9 lists the regression analysis result for machine A. The t-statistic is used to test if each predictor is significant in the regression model, the test is conducted below:

$t_{constant} = 2.99 > t_{(0.025,6)} = 2.447$, the constant is significant in the model.

$t_{Defect\ Time} = 14.53 > t_{(0.025,6)} = 2.447$, the defect time is significant in the model.

$t_{DV1} = 0.37 < t_{(0.025,6)} = 2.447$, the variable DV1 is not significant in the model.

$t_{DV2} = 0.39 < t_{(0.025,6)} = 2.447$, the variable DV2 is not significant in the model.

$t_{DV3} = 0.77 < t_{(0.025,6)} = 2.447$, the variable DV3 is not significant in the model.

As the variables DV1, DV2, and DV3 are not significant in the model, it means that the type of failure is not significant for this model; hence, we can remove the failure type from the model and repeat the regression again.

Regression Result (second run, after removing DV1, DV2, and DV3 from the model):

The second run of regression analysis shows a high value of R-square. Table 11.10 lists the standard error and R-square values for the second run of regression analysis.

Table 11.11 lists the ANOVA table for the second run of regression analysis with F statistic equals to 309.34, which is higher than the F table critical value (5.12) at $\alpha = 0.05$. The F test shows that the model is significant.

$F = 309.34 > F_{(0.05,1,9)} = 5.12$; the model is significant.

Table 11.10 Standard Error and *r*-square Values for the Second Run of the Regression Analysis of Machine A

Standard Error (σ)	919.170
R-Square	97.2%
R-Square (adj)	96.9%

Table 11.11 The ANOVA Table for the Second Run of the Regression Analysis Machine A

Source	DF	SS	MS	F	P
Regression	1	261351583	261351583	309.34	0
Residual error	9	7603867	844874		
Lack of fit	1	4178587	4178587	9.76	0.014
Pure error	8	3425280	428160		
Total	10	268955450			

$F_{\text{lack of fit}} = 9.76 > F_{(0.05,1,8)} = 5.32$; the model shows evidence of lack of fit. According to the *R*-square value and the fitted line plot that is illustrated in Figure 11.4, we cannot reject the model and will consider it significant.

Table 11.12 lists the results of the second run of regression analysis. The constant and the defect time variable are significant in the model.

$t_{\text{constant}} = 5.47 > t_{(0.025,9)} = 2.262$, the constant is significant in the model.

$t_{\text{Defect Time}} = 17.59 > t_{(0.025,9)} = 2.262$, the defect time is significant in the model.

The Regression Model

The regression model used is stated as:

$$\text{Failure Time} = 2445.6 + 0.8603 * \text{Defect Time}$$

$$Y = 2445.6 + 0.8603\, X_1$$

The residuals versus failure time plot for regression of machine A is illustrated in Figure 11.5, and the normal probability plot of residuals for regression of machine A is illustrated in Figure 11.6.

Figure 11.5 illustrates the residuals in three groups; each group contains points that are aligned. Normally, this is not considered as random behavior of residuals as they should be distributed randomly around the zero line. The aligning of the

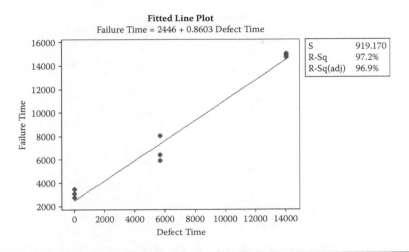

Figure 11.4 The fitted line plot for the second run of regression analysis of machine A.

Table 11.12 The Results of the Second Run of Regression Analysis of Machine A

Predictor	Coefficient	Standard Error (SE)	t-Statistic	p-Value
Constant	2445.6	447.5	5.47	0
Defect time	0.8603	0.04891	17.59	0

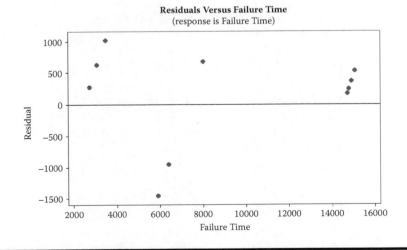

Figure 11.5 The residuals versus failure time for regression of machine A.

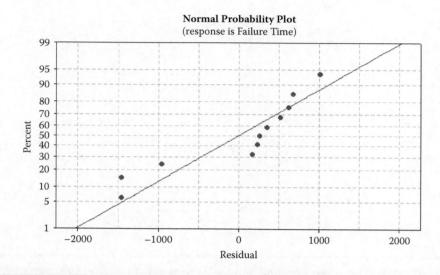

Figure 11.6   The normal probability plot of residuals.

points is due to repetition of the independent variable values (defect time) in the original regression data.

The regression model that is used entails the assumption of residuals normality; hence to accept a regression model, residuals should be normally distributed. Normality test has been conducted for the residuals; the result of Kolmogorov–Smirnov test is shown below:

H_0: The residuals are distributed normally with $\mu = 0.0$ and $\sigma = 872.0$.
H_1: The residuals are not distributed normally.
D statistic $= 0.303$.
At $\alpha = 0.05$, D *critical* $= 0.249$.

As D statistic $> D$ critical, the null hypothesis cannot be accepted, the residuals are not normally distributed. Transformation is required for the original data (for the failure times only) and regression analysis is repeated again. Normality can be attained using the Box–Cox transformation. The new transformed data points are given by

$$Y_{new} = \sqrt{y}$$

Any prediction from the new regression model after transformation should be updated to the original scale of the data, which means the predicted value should be squared to match the original data scale.

Regression Analysis after Data Transformation

The transformed data have been used for regression analysis. R-square value is considered a high value that represents a good regression model. Table 11.13 lists the standard error and *R*-square values.

Table 11.14 lists the ANOVA table with F statistic equals to 509.18 which is higher than the *F* table critical value (5.12) at $\alpha = 0.05$. The F test shows that the model is significant.

$F = 509.18 > F_{(0.05,1,9)} = 5.12$; the model is significant.

Table 11.15 lists the regression analysis result for machine A after transformation. The *t*-statistic is used to test whether each predictor is significant in the regression model; the test is conducted below.

$t_{constant} = 28.34 > t_{(0.025,9)} = 2.262$, the constant is significant in the model.
$t_{Defect\ Time} = 22.57 > t_{(0.025,9)} = 2.262$, the defect time is significant in the model.

Table 11.13 Standard Error and R-Square Values for Regression Analysis of Machine A after Transformation

Standard error (σ)	3.960
R-Square	98.3%
R-Square (adjusted)	98.1%

Table 11.14 The ANOVA Table of Regression Analysis of Machine A after Data Transformation

Source	DF	SS	MS	F	P
Regression	1	7987.9	7987.9	509.18	0.000
Residual error	9	141.2	15.7		
Total	10	8129.1			

Table 11.15 The Regression Analysis Result for Machine A After Data Transformation

Predictor	Coefficient	Standard Error (SE)	t-Statistic	p-Value
Constant	54.64	1.928	28.34	0.00
Defect time	0.00475	0.00021	22.57	0.00

The Regression Model

$$(\text{Failure Time})^{0.5} = 54.64 + 0.0047 * \text{Defect Time}$$

$$Y_{new}^{0.5} = 54.64 + 0.0047\ X_1.$$

The fitted line plot with confidence bounds for regression of machine A after data transformation is illustrated in Figure 11.7.

The residuals versus failure time plot for regression of machine A after data transformation is illustrated in Figure 11.8, and the normal probability plot of residuals for regression of machine A after transformation is illustrated in Figure 11.9.

The residuals that are illustrated in Figure 11.9 appear in three groups due to the repetition in the original and transformed data. The residuals can be considered randomly distributed.

The results of Kolmogorov–Smirnov test for normality is:

$$D \text{ statistic} = 0.189$$

For sample size $n = 11$, and $\alpha = 0.05$; $D\ critical = 0.249$.

The residuals can be considered normally distributed; hence, the regression model is significant.

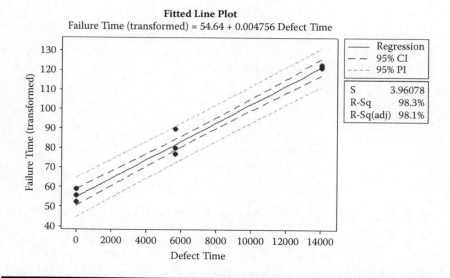

Figure 11.7   The fitted line plot with confidence bounds for regression of machine A after data transformation.

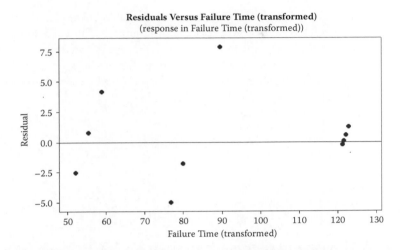

Figure 11.8   The residuals versus failure time plot for regression of machine A after data transformation

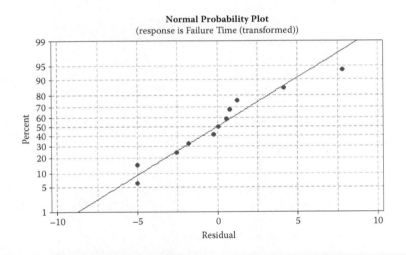

Figure 11.9   The normal probability plot of residuals for regression of machine A after transformation.

Regression Diagnostics

A regression model should be validated before it is used for future predictions. This test ensures that the assumptions of the regression model are reasonable and that data appear to be sampled from a population that meets the assumptions. The main assumption is the linearity of residuals, that is, the difference between the observed and the predicted values of a dependent variable. Regression diagnostics check the fitted model for undesirable residuals and outliers. This checking process may suggest a reformulation of the model that will improve the prediction accuracy.

The residuals plot may show outliers; these are residuals with an extremely high value, either positive or negative. One common test for outliers detection is using studentized residuals. The test statistic, d_i, is given by

$$d_i = \frac{e_i}{\sqrt{MSE_i}\sqrt{1-h_{ii}}} \tag{11.1}$$

where

e_i is the i residual
MSE_i is the MSE when observation i is deleted
h_{ii} is the ith diagonal element of the hat matrix.

For regression in the matrix notation, the hat matrix $H = X(X'X)^{-1}X'$, X is the matrix of the independent variables, and X' is the transpose of matrix X.

The hypothesis test for an outlier is:

H_o: observation i is not an outlier.

H_l: observation i is an outlier.

A point is an outlier, therefore, and dropped from the regression model if H_o is true, that is, if d_i is in the extreme upper or lower tail of the t-distribution at a level of significance α and a degree of freedom $(n - k - 2)$, where n is the number of observations and k is the number of parameters in the regression model.

If an outlier is detected, the data point that corresponds to this outlier may be deleted, and the regression should be repeated. The new regression model will be more accurate (lower MSE and higher R^2). An outlier can only be deleted if it is not at a high leverage point or not highly influential.

A high leverage point is an observation that has a substantial impact on the regression model due to its differences from other observations on one or more of the independent variables. Leverage h_{ii} is the ith diagonal element of the hat matrix. The rule-of-thumb to classify leverage as high is a leverage value of greater than $2(k+1)/n$.

An observation is influential if its deletion changes various estimates drastically. Influential observations may have substantial differences from other cases of the independent variables, extreme values, or both at the same time. Cook's distance is one measurement of the influence of an observation, given by

$$D_i = \frac{e_i^2 h_{ii}}{(K+1)MSE(1-h_{ii})^2} \tag{11.2}$$

Cook's distance measures the change in all other residuals when the observation i is deleted. D_i follows an F distribution if $D_i < F(0.2; k + 1, n - k - 1)$; then the ith observation has little influence. If $D_i > F(0.5; k + 1, n - k - 1)$, then the ith observation has high influence. Otherwise, the observation is somewhat influential. An observation having little influence can be deleted if it is an outlier, if it does not have high leverage, so as to improve the regression model. On the other hand, an observation of high influence should not be deleted if it is an outlier, as it will affect the regression model. For outlier observations that are between low and high influence, deletion decisions depend on the leverage value and the regression model.

The reasons for outliers should be investigated as to the possible lowest level, even to the level of data collection if necessary. It is preferable that an observation be replaced with a correct value if an error appears in data collection or entry, rather than deleting the observation. A justification for an outlier that occurred should be stated before an outlier is deleted.

Based on the regression diagnostics, the regression analysis may be repeated and the final regression model developed after going through different tests to determine the best model to represent the relationships among the variables and to predict more accurately future outcomes. The diagnostic should be done once; analysis should not go into repeated loops of model checking. Regression diagnostic was conducted for each regression model analysis.

The randomness and normality of residuals have been tested before for machine A, and they are accepted. Other diagnostic tests such as outliers, leverage, and influence are shown below.

The Current Regression Model (n = 11, k = 1):

$$Y_{new}^{0.5} = 54.64 + 0.0047\, X_1$$

$R^2 = 98.3\%$ $\qquad$ $SSE = 141.2$ $\qquad$ $MSE = 15.7$ $\qquad$ $PRESS = 186.04$

Outliers Detection:

Vector d_i shows the studentized (deleted) residuals; the p-values are shown for each case according to the critical t-value.

$$
d_i = \begin{bmatrix} 0.214 \\ 1.232 \\ -0.720 \\ -0.455 \\ -1.401 \\ \mathbf{2.728} \\ -1.401 \\ 0.149 \\ 0.336 \\ -0.064 \\ 0.015 \end{bmatrix}, \quad t_{0.05,\,11\text{-}1\text{-}2} = t_{0.05,\,8} = 1.86, \quad P-values = \begin{bmatrix} 0.417 \\ 0.126 \\ 0.246 \\ 0.330 \\ 0.099 \\ \mathbf{0.0129} \\ 0.099 \\ 0.442 \\ 0.372 \\ 0.475 \\ 0.493 \end{bmatrix}
$$

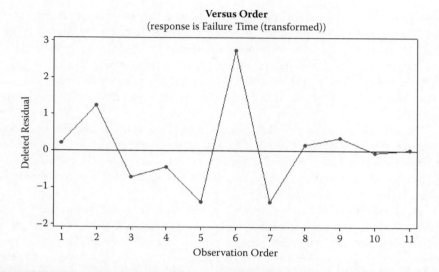

Figure 11.10 Deleted residuals versus order for machine A.

The null hypothesis cannot be rejected for all observations except observation number 6 where p-value = 0.0129, therefore, observation number 6 is an outlier. Before repeating the regression, leverage and Cook's distance should be tested. Figure 11.10 illustrates deleted residuals versus order for machine A. It is obvious that observation number 6 is an outlier with respect to other observations.

Leverage

Vector L shows the leverage value for regression model of machine A. According to the rule-of-thumb, $2(1 + k)/n = 2(1 + 1)/11 = 0.3636$, a leverage is considered high if it is above 0.3636 for the given model. There is no high leverage observation for this model.

$$
L_i = \begin{bmatrix}
0.237 \\
0.237 \\
0.237 \\
0.097 \\
0.097 \\
0.097 \\
0.097 \\
0.225 \\
0.225 \\
0.225 \\
0.225
\end{bmatrix}
$$

Cook's Distance

Vector D_i shows the Cook's distance for the observations of the given model.

$$
D_i = \begin{bmatrix}
0.008 \\
0.223 \\
0.085 \\
0.012 \\
0.095 \\
0.233 \\
0.095 \\
0.003 \\
0.018 \\
0.000 \\
0.000
\end{bmatrix}
$$

According to the F-distribution:

$$
F_1 = F(0.2;\, k + 1,\, n - (k + 1)) = F(0.2;\, 2,\, 9) = 0.228
$$
$$
F_2 = F(0.5;\, k + 1,\, n - (k + 1)) = F(0.5;\, 2,\, 9) = 0.749
$$

All observations have little influence except for observation number 6 that is slightly above F_1, and as it is an outlier, this observation can be deleted without high influence on the regression model. Regression is repeated after deletion of observation number 6.

Regression after Deletion of Outlier Point (observation number 6):

The regression is repeated for $n = 10$, and the new model is:

$$
Y_{new}^{0.5} = 53.6 + 0.0047 X_1
$$

$R^2 = 99.1\%$ $SSE = 73.1$ $MSE = 9.1$ $PRESS = 111.64$

This model is better than the previous model, as the SSE, MSE, and $PRESS$ values have become much lower than the values in the previous model. As the model has changed, the confidence intervals of the coefficients and the predicted value should be updated.

It is observed that the type of failure in general is not significant in the regression models. Table 11.16 lists a summary of the regression models for each machine.

Table 11.16 Summary of the Regression Models for Each Machine

Machine	Regression Model	Variables
Machine A	$Y_{new}^{0.5} = 53.6 + 0.0047\, X_1$	X_1 : Defect time
Machine B	$Y = 84.8 + X_1$	X_1 : Defect time
Machine C	$Y = 322 + X_1$	X_1 : Defect time
Machine D	$Y = 127 + X_1$	X_1 : Defect time
Machine E	$Y = 137 + X_1$	X_1 : Defect time
Machine F	$Y = X_1 + 485\, X_2$	X_1 : Defect Time $X_2 = 1$ for hydraulic failure, 0 for other failures

11.9 Stage 3: System Simulation

As the regression model for each machine was validated, the predicted failure times were used to set up a schedule for maintenance activities. This schedule addressed the production schedule so that the maintenance activity for the predicted failures could be performed without interruption, thus reducing breakdowns. The schedule of maintenance activities was based on the predicted failure times from defects due to machine malfunction. The defect times were generated by a simulation model using historical defect times.

11.9.1 Current System Simulation Model

A simulation model that addressed the current situation of the system was developed; this model was called the as-is model. This model did not address defect times; it considered machines in the production line with the random failure and repair distributions developed in the distribution fitting step. This model represented the current state of the system without any experimentation or modification. When this model was verified and validated, defect-times generation were incorporated to predict the failure of each machine. The as-is simulation model was built using Witness 2010 simulation software, a powerful simulation software used worldwide by manufacturing and service organizations.

Simulation starts by the generation of material parts for the first machine (machine A) to a buffer. Once a part arrives to machine A, processing starts and lasts for the processing time associated with this machine. The failures are modeled to occur while the machine is busy, as Witness has many options for the failure to occur. This represents the actual manner for failure to occur. An assumption holds that if a breakdown occurs while the machine is processing a part, this part is not considered to be a defect; it will remain in the line until final production. This assumption

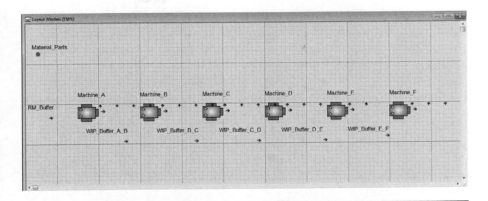

Figure 11.11 The as-is Witness simulation model elements.

follows what actually happens in the real system. When a part is finished processing by machine A, it is sent to a buffer between machine A and machine B.

Once machine B is available, it will start processing parts from a buffer between machine A and machine B and sends them to a buffer between machine B and machine C. There is a buffer between every two machines. The output of machine F is sent to a warehouse that Witness uses as a finished product warehouse. Random failures and repairs occur for each type of failure for each machine while a machine is processing according to the fitted statistical distributions. Figure 11.11 illustrates the as-is Witness simulation model elements.

The time that the production line was on and off was modeled in Witness. A shift pattern of on six days a week, 16 hours a day was developed and incorporated in each element of the model, including the arrival of the raw materials. The 16 working hours form shift one and shift two (eight hours each), while shift three (eight hours) is a rest time. Shift one starts at 8:00 and ends at 16:00 every working day, while shift two starts at 16:00 and ends at 00:00 every working day. The rest shift starts at 00:00 and ends at 8:00. The unit of time that used is in hours, which matches the fitted statistical distributions of failure and repair times.

11.9.1.1 Simulation Verification

In simulation verification, a model is checked against the developer's blueprint. Verification is addressed by checking the details of each element in the model, for example, by checking the input and output for each machine, the processing time, and the breakdown and repair times.

11.9.1.2 Simulation Validation

Validation is accomplished for the purpose of developing a model that represents a real system in every detail. There is no unique validation process; essentially it considers a model's output for different levels and compares it statistically with a real system's output. A validated model is known as a model that passes certain statistical tests that give the model the credibility to be used for further experimentations. For the as-is model that was developed, validation was accomplished for system's throughput and random failure and repair statistical distribution. The throughput of the simulation must be similar as the system's throughput. On average, the given production line has a throughput of 4 to 6 items daily. The simulation ran for 672 hours (4 weeks * 168 hours per week) with a throughput of 85 items giving an average of 21.25 items per week and 3.54 items per day, which is below the system's average. The model must reach steady state where, for some machines, TBF is more than 168 hours. The throughput is validated through another two runs of 12 weeks (2016 hours) and 30 weeks (5040 hours), where the throughputs are 380 items and 1004 items that is an average of 31.66 and 33.46 items per week and 5.27 items and 5.57 items per day, respectively. This shows that the model's throughput is validated.

Before validation of random failure and repair times, machine statistics should be reviewed for validation of performance. The statistics are the percentages of idle, busy, blocked, and broken times. Table 11.17 lists machine statistics after a run of 30 weeks (5040 hours). The relative high percentages of blocked time for machine A to machine D are due to high percentage of broken time for machine E as machines are in series. Also, machine E is the slowest in the line; this increases the block time for machines preceding machine E and increases the idle time for machine F.

Efficiency is defined as percentage of busy time with respect to available time; overall efficiency of the line, therefore, is 69.79%, the lowest machine efficiency (machine D). The actual system has efficiency between 70% and 75%, which is approximately similar to the efficiency of the simulation model. Historical efficiency data are not available for each machine, so validation is performed for efficiency of the overall system.

Validation is accomplished for random failure and repair times. This entails validating random failure and repair times that are modeled for each machine

Table 11.17 Machines Statistics after a Run of 30 Weeks (5040 hours)

Machine	Idle Time (%)	Busy Time (%)	Blocked Time (%)	Broken Time (%)
Machine A	0.0	70.96	27.77	1.27
Machine B	0.07	79.36	18.49	2.07
Machine C	0.80	70.21	27.73	1.27
Machine D	0.53	69.79	24.58	5.09
Machine E	1.19	86.84	0.32	11.65
Machine F	19.74	78.08	0.00	2.18

according to historical statistical distribution. Validation is the comparison of two sets of data with the hypothesis that they belong to the same statistical distribution. For example, machine A has four types of failure with corresponding repair times. The validation at this step compares a generated set of data from the simulation model to the historical data of the same type of failure.

A user-defined code was developed for the simulation model to generate data point and sent to external file that contains the time of certain events in the simulation. These events are the time when a machine breakdown (failure time) and the time when the machine is back on (time of repair); this was coded for each machine for each type of failure. Then, using this generated data, TBF and repair times RT are calculated. Generated and historical TBF and RT are used for validation.

For statistical analysis a plot is developed for mechanical type generated TBF and mechanical type historical TBF. The objective is to validate whether the generated set of data comes from the same distribution of the historical set of data. The plot is for the quantiles of the data.

11.10 Modified Simulation Model

The as-is model has been validated. Changes can be made on this model to experiment and track results on the real system. A new model called the defect-time model was developed. This new model was developed to generate predicted failure times (PFT) from generated defect times (DT) so as to develop a maintenance plan that is synchronized with production. This maintenance plan addresses the predicted failures and recommends a schedule for preventive maintenance action to avoid breakdown of the system, hence to increase the available time for production.

PFT is generated using the regression model for each machine. A failure is predicted if there is a product defect due to machine malfunction. The "defect time" simulation model generates a defect time for each machine that is used in the regression model for the prediction of a failure. The generation of a defect time is based on the historical time between defects (TBD) for each machine.

It is not always possible to generate a defect time in the simulation model according to the exact value of TBD because at this exact time a machine could be idle (waiting for parts), broken, or being repaired. A range of ± 10% of TBD (0.9*TBD, 1.1*TBD) has been assumed for the generation of defect times. A defect can occur in this range at any time while the machine is processing a part.

As the processing time of machines in the case study is very short with respect to the TBD, a DT is generated whenever the time of a part to leave a machine is within the set range of TBD. The simulation model is forced to only consider the time of the first defect event even if there is more than one that can be generated within the set range of TBD; this is as defects occur in reality every TBD time.

When run time exceeds (1.1*TBD), the next DT is generated in the range of the time of (2*0.9*TBD, 2*1.1*TBD). The range of TBD for the next cycle when run time exceeds (2*1.1*TBD) is (3*0.9*TBD, 3*1.1*TBD). This range is updated to address the run time to be able to generate DT as the system runs.

The generation of DT has been coded for each machine in the case study, so a failure can be predicted for each machine. A code can be run as an action in Witness for a machine element when a machine finishes processing a job for a part. As TBD is relatively long, only four TBD ranges are needed to develop a plan for about four months to three years. The four ranges are given by (X*0.9*TBD, X*1.1*TBD), where X = 1, 2, 3, and 4. These ranges are for each machine according to its TBD. Four real variables are defined for each machine. Each variable holds the DT within a corresponding range; initially each variable has a value of zero, but each will be updated later as simulation runs and assigns a value for this variable. For example, variable "DT_A_1" represents the defect time for machine A in the first range.

Using the IF conditional statement, a code is set to generate a DT for a certain machine in a certain range of TBD and stamp this value for the corresponding DT variable. The IF statement is true if the time that a part leaves a machine is within one of the ranges of TBD, then the corresponding DT variable will have a value generated using the corresponding range of TBD; otherwise, no action is performed. For example, for machine C, if the time a part leaves the machine is between 0.9*TBD = 2134 and 1.1*TBD = 2608, the variable DT_C_1 has a value in the range of (2134, 2608); otherwise, it remains zero. The range of TBD is longer than the processing time of a machine, so there will be a defect time generated for each TBD range.

To ensure that the simulation model assigns one value for each DT variable, another four binary variables are defined. Each binary variable is related to a DT variable. The initial value of each binary variable is zero. The IF statement used to assign a value for each DT variable holds only if the corresponding binary variable equals zero. Once a value is assigned to a DT variable, the corresponding binary variable is updated to one; so the IF statement will no longer change the DT variable. The variables are displayed on the screen and are exported to external files that can be read by other software.

This coding has been done for each machine in the system, so the DT for each machine can be used to calculate PFT using the corresponding regression model. A total of four variables that represent the PFT are defined for each machine except machine F, as the type of failure is significant in the regression model for this machine. Failures can be predicted as hydraulic failure (when the variable of the type of failure equals one) or other type of failure (when the variable of the type of failure equals zero). The PFT variables are displayed on the screen, and are exported to external files that can be read by other software. This routine enables the PFT to be found for each machine and a maintenance schedule to be developed that is synchronized with production. All codes that are used in Witness are shown in the appendix.

11.11 Stage 4: Maintenance Schedule

Preventive maintenance is essential in maintaining or increasing production time; further, maintenance and production should be synchronized. If a random failure occurs for a certain machine in the system, production is stopped until the machine is repaired and available for production, a stoppage that can cost an organization a large amount of money. On the other hand, if a failure is predicted to occur at certain time, preventive maintenance action can help avoid the breakdown. Generally, the preventive maintenance actions or jobs are performed during rest hours in the shift. This plan will not interrupt production, as machines are already idle. This section provides a schedule for preventive maintenance jobs according to the PFT for each machine.

Table 11.18 lists the PFTs and the successive random failure times (using TBF distributions) for each machine for each type of failure generated using the "defect time" simulation model. Four failure times are predicted for each machine from four corresponding generated defect times, including the predicted failures for about three years. For example, a defect time for machine B is generated at 4569 hours, a failure is predicted at 4654 hours. The table also lists the time for the next random failure after the predicted failure at 4654 hours for each type of failure for machine B. There is a mechanical, electrical, hydraulic, and coolant random failure at 4693, 4995, 4671, and 5007 hours, respectively, that occur after the time of the predicted failure at 4654 hours. There are no failures between the PFT at 4654 hours and the next different types of random failure occurrences. If preventive maintenance addresses the PFT, it avoids breakdown at the predicted time during production, and at the same time, avoids random failures.

Table 11.19 lists time between predicted failure and the predecessor random failure for each type of failure. Predecessor random failure is used to calculate the probability of a predicted failure, which is also listed in the table. The PFTs and predecessor random failure times are generated using the "defect time" simulation model. As the PFT does not identify the type of failure that is predicted to occur, a probability of occurrence for each type of failure is calculated using time between predicted failure and the predecessor random failure through the TBF cumulative distribution function that is fitted for each machine for each type of failure. For example, the probability of the predicted failure at 2456 hours for machine C is 0.9740, 0.8894, 0.8894, and 0.9586 for mechanical, electrical, hydraulic, and coolant failure, respectively. The failure with highest probability is considered for maintenance scheduling. Highest probabilities of failure for each PFT for each machine are shown in the shaded cells in Table 11.19. The regression model for machine F predicts a failure to be either hydraulic or other type, so hydraulic failures are considered for planning regardless of their probability, but other failures for machine F (mechanical or electrical) that have the higher probability of occurrence are considered for maintenance planning.

Table 11.20 lists date, time, and probability of predicted failures. Simulation runs at 01/03/2011–08:00 starting time. A schedule for preventive maintenance

Table 11.18 The PFTs and the Successive Failure Times for Each Machine for Each Type of Failure

M/C	Generated DT (hr)	PFT (hr)	Time of the Successive Random Failure after the Predicted Failure (hr)				Difference between Time of the Successive Random Failure after the Predicted Failure and the PFT (hr)			
			Mech.	Elec.	Hyd.	Coolant	Mech.	Elec.	Hyd.	Coolant
A	6330	6947	8851	7798	105397	7500	1904	851	98450	553
	12658	12790	13262	13329	105397	13823	472	539	92607	1033
	19016	20443	22095	23437	105397	20871	1652	2994	84954	428
	25316	29786	30040	30454	105397	31847	254	668	75611	2061
B	4569	4654	4693	4955	4671	5007	39	301	17	353
	9115	9200	9348	9393	11079	9907	148	193	1879	707
	13712	13797	13823	14217	14072	16415	26	420	275	2618
	18230	18315	18371	18396	18935	19822	56	81	620	1507
C	2134	2456	2781	3181	6497	5628	325	725	4041	3172
	4269	4591	4620	4771	6497	5628	29	180	1906	1037
	6416	6738	6778	8658	8390	9375	40	1920	1652	2637
	8655	8977	8994	10747	22295	9375	17	1770	13318	398

D	2410		2537	3659	2753	–	2584	1122	216	–	47
	4816		4943	5024	4959	–	6854	81	16	–	1911
	7257		7384	7764	10209	–	8699	380	2825	–	1315
	9634		9761	9900	10209	–	9886	139	448	–	125
E	2433		2570	3158	3668	–	–	588	1098	–	–
	4858		4995	5075	5083	–	–	80	88	–	–
	7293		7430	8009	7551	–	–	579	121	–	–
	9706		9843	9910	9944	–	–	67	101	–	–
F	4064	Hyd.	4549	–	–	5820	–	–	–	1271	–
		Other	4064	4459	4366	–	–	413	320	–	–
	8096	Hyd.	8581	–	–	10284	–	–	–	1703	–
		Other	8096	9040	8384	–	–	944	288	–	–
	12140	Hyd.	12625	–	–	14054	–	–	–	1429	–
		Other	12140	12596	12398	–	–	456	258	–	–
	16186	Hyd.	16671	–	–	16956	–	–	–	285	–
		Other	16186	16555	16407	–	–	369	221	–	–

Table 11.19 Time between Predicted Failure and the Predecessor Random Failure for Each Type of Failure and the Probability of the Predicted Failure

M\C	Time between Predicted Failure and the Predecessor Random Failure (hr)				Probability of Predicted Failure			
	Mech.	Elec.	Hyd.	Coolant	Mech.	Elec.	Hyd.	Coolant
A	1456	1405	6066	1963	**0.9753**	0.9517	0.9529	0.9500
	264	2243	11909	2407	0.7455	**0.9814**	0.9775	0.9658
	2987	124	19562	2471	**0.9964**	0.6385	0.9883	0.9676
	255	2308	28905	459	0.7384	0.9827	**0.9936**	0.7625
B	28	126	36	623	0.6328	0.7617	0.2069	**0.8713**
	24	404	1263	1480	0.5965	0.9519	**0.9903**	0.9446
	84	204	156	367	**0.8681**	0.8584	0.5492	0.8100
	232	269	185	281	**0.9795**	0.9032	0.6014	0.7754
C	736	406	2444	2456	**0.9740**	0.8894	0.8894	0.9586
	1027	190	4579	4591	0.9899	0.7861	0.9387	**0.9923**
	1493	1304	239	265	**0.9976**	0.9778	0.5992	0.6112
	76	151	586	2504	0.6330	0.7496	0.7229	**0.9602**
D	20	200	–	19	0.4627	**0.8621**	–	0.4992
	129	295	–	495	0.8236	0.9231	–	**0.9526**
	248	1421	–	245	0.9169	**0.9998**	–	0.8733
	272	3798	–	25	0.9270	**1.0000**	–	0.5345
E	252	437	–	–	0.9148	**0.9303**	–	–
	1042	32	–	–	**0.9989**	0.6120	–	–
	622	68	–	–	**0.9899**	0.7048	–	–
	31	467	–	–	0.5311	**0.9363**	–	–
F	773	115	197	–	**0.9989**	0.7389	**0.6581**	–
	240	425	1960	–	**0.9570**	0.9480	**0.9594**	–
	628	53	2341	–	**0.9971**	0.5918	**0.9716**	–
	160	312	1076	–	**0.9180**	0.9110	**0.9001**	–

Table 11.20   Date, Time, and Probability of Predicted Failures

M/C	PFT (hr)		Type of Failure	Probability	Date and Time of Predicted Failure (MM/DD/YYYY HH:MM)
A	6947		Mechanical	0.9753	10/19/2011 11:00
	12790		Electrical	0.9814	06/18/2012 22:00
	20443		Mechanical	0.9964	05/03/2013 19:00
	29786		Hydraulic	0.9936	05/28/2014 02:00
B	4654		Coolant	0.8713	07/15/2011 22:00
	9200		Hydraulic	0.9903	01/21/2012 08:00
	13797		Mechanical	0.8681	07/30/2012 21:00
	18315		Mechanical	0.9795	02/04/2013 03:00
C	2456		Mechanical	0.9740	04/15/2011 08:00
	4591		Coolant	0.9923	07/13/2011 07:00
	6738		Mechanical	0.9976	10/10/2011 18:00
	8977		Coolant	0.9602	01/12/2012 01:00
D	2537		Electrical	0.8621	04/18/2011 17:00
	4943		Coolant	0.9526	07/27/2011 23:00
	7384		Electrical	0.9998	11/06/2011 16:00
	9761		Electrical	1.0000	02/13/2012 17:00
E	2570		Electrical	0.9303	04/20/2011 02:00
	4995		Mechanical	0.9989	07/30/2011 03:00
	7430		Mechanical	0.9899	11/08/2011 14:00
	9843		Electrical	0.9363	02/17/2012 03:00
F	Hyd.	4549	Hydraulic	0.6581	07/11/2011 13:00
	Other	4064	Mechanical	0.9989	06/21/2011 08:00
	Hyd.	8581	Hydraulic	0.9694	12/26/2011 13:00
	Other	8096	Mechanical	0.9570	12/06/2011 08:00
	Hyd.	12625	Hydraulic	0.9716	06/12/2012 01:00
	Other	12140	Mechanical	0.9971	05/22/2012 20:00
	Hyd.	16671	Hydraulic	0.9001	11/27/2012 15:00
	Other	16186	Mechanical	0.9180	11/07/2012 10:00

jobs was developed taking into account the dates and times in Table 11.20. As there are three shifts of eight hours each, and two of these shifts are working times, preventive maintenance jobs are scheduled to be performed on the third shift (rest shift). Working starts at 08:00, the first shift ends at 16:00, as the second shift starts. The second working shift ends at 00:00. The rest shift is from 00:00 to 08:00.

The times of day for some predicted failures are not within the working shift time, but the generated DT is while a machine is processing, which matches the real scenario. PFTs are generated using regression models, so a PFT depends on the regression model used, and it will be different for a production line that works different hours; a failure can occur at any time. The majority of failures occur while a machine is processing, however, there is a probability that a previously functioning machine or a system will fail while it is not working. For example, a machine might not work at the beginning of a shift after a rest shift; yet there might have been some sign from this machine of a coming breakdown. In an instance such as this, condition based maintenance is said to be involved.

For the given PFT, a failure is considered to occur at the beginning of the working shift if PFT is on the rest shift, because the machine was working properly in the working shift prior to the rest shift, yet is not available at the beginning of the working shift. PFTs on the rest shift are shifted forward to the next working shift. Table 11.21 lists PFTs and shifted PFTs for PFTs on the rest shift for each machine. Scheduling is developed for shifted PFTs. When a failure is predicted to occur in a working shift, a preventive maintenance job is planned in the preceding rest shift. According to the predicted failure type, preventive maintenance action addresses the corresponding system to the failure type. This action prevents breakdowns for the production line due to sudden failures.

A policy can be developed to check the different systems for each machine on the rest shift; there will be no need for a schedule and no need to predict a failure as preventive maintenance jobs for the entire system are performed. But this policy is not reasonable and not cost effective, because each machine in a system should be checked for all possible failures, which is time and effort consuming. Preventive maintenance jobs, according to PFTs, check a certain system for critical and possible causes that may breakdown a machine. Thus, a system's availability is increased, and so the overall cost is reduced.

A predicted failure does not occur exactly at PFT, as the system is not deterministic in terms of failures. A preventive maintenance job can be performed at a time close to PFT to avoid the occurrence of the failure. Table 11.22 lists the schedule of preventive maintenance jobs according to the shifted PFTs column shown in Table 11.21. The preventive maintenance is performed in the rest shift that starts from 00:00 to 08:00. The proposed maintenance schedule shows the time a job is to be performed (starting time), the machine that needs maintenance, and the system to check for possible failure. The next preventive maintenance job to be performed is on 04/15/2011 for machine C for the mechanical system. The last preventive maintenance job that this schedule shows is on 05/28/2014 for machine A for the hydraulic system.

Table 11.21 PFT and Shifted PFT's for PFT on the Rest Shift for Each Machine

M/C	PFT (hr)	Type of Failure	Probability	Date and Time of Predicted Failure (MM/DD/YYYY HH:MM)	Shifted date and Time of Predicted Failure (MM/DD/YYYY HH:MM)
A	6947	Mechanical	0.9753	10/19/2011 11:00	10/19/2011 11:00
	12790	Electrical	0.9814	06/18/2012 22:00	06/18/2012 22:00
	20443	Mechanical	0.9964	05/03/2013 19:00	05/03/2013 19:00
	29786	Hydraulic	0.9936	**05/28/2014 02:00**	**05/28/2014 08:00**
B	4654	Coolant	0.8713	07/15/2011 22:00	07/15/2011 22:00
	9200	Hydraulic	0.9903	01/21/2012 08:00	01/21/2012 08:00
	13797	Mechanical	0.8681	07/30/2012 21:00	07/30/2012 21:00
	18315	Mechanical	0.9795	**02/04/2013 03:00**	**02/04/2013 08:00**
C	2456	Mechanical	0.9740	04/15/2011 08:00	04/15/2011 08:00
	4591	Coolant	0.9923	**07/13/2011 07:00**	**07/13/2011 08:00**
	6738	Mechanical	0.9976	10/10/2011 18:00	10/10/2011 18:00
	8977	Coolant	0.9602	**01/12/2012 01:00**	**01/12/2012 08:00**
D	2537	Electrical	0.8621	04/18/2011 17:00	04/18/2011 17:00
	4943	Coolant	0.9526	07/27/2011 23:00	07/27/2011 23:00

(Continued)

Table 11.21 (*Continued*)   PFT and Shifted PFT's for PFT on the Rest Shift for Each Machine

M/C	PFT (hr)		Type of Failure	Probability	Date and Time of Predicted Failure (MM/DD/YYYY HH:MM)	Shifted date and Time of Predicted Failure (MM/DD/YYYY HH:MM)
	7384		Electrical	0.9998	11/06/2011 16:00	11/06/2011 16:00
	9761		Electrical	1.0000	02/13/2012 17:00	02/13/2012 17:00
E	2570		Electrical	0.9303	**04/20/2011 02:00**	**04/20/2011 08:00**
	4995		Mechanical	0.9989	**07/30/2011 03:00**	**07/30/2011 08:00**
	7430		Mechanical	0.9899	11/08/2011 14:00	11/08/2011 14:00
	9843		Electrical	0.9363	**02/17/2012 03:00**	**02/17/2012 08:00**
F	Hyd.	4549	Hydraulic	0.6581	07/11/2011 13:00	07/11/2011 13:00
	Other	4064	Mechanical	0.9989	06/21/2011 08:00	06/21/2011 08:00
	Hyd.	8581	Hydraulic	0.9694	12/26/2011 13:00	12/26/2011 13:00
	Other	8096	Mechanical	0.9570	12/06/2011 08:00	12/06/2011 08:00
	Hyd.	12625	Hydraulic	0.9716	**06/12/2012 01:00**	**06/12/2012 08:00**
	Other	12140	Mechanical	0.9971	05/22/2012 20:00	05/22/2012 20:00
	Hyd.	16671	Hydraulic	0.9001	11/27/2012 15:00	11/27/2012 15:00
	Other	16186	Mechanical	0.9180	11/07/2012 10:00	11/07/2012 10:00

Table 11.22 Preventive Maintenance Schedule for the Predicted Failures

M/C	Type of Failure	Shifted Date and Time of Predicted Failure (MM/DD/YYYY HH:MM)	Scheduled Starting Time for Preventive Maintenance Job
Machine C	Mechanical	04/15/2011 8:00	04/15/2011 00:00
Machine D	Electrical	04/18/2011 17:00	04/18/2011 00:00
Machine E	Electrical	04/20/2011 8:00	04/20/2011 00:00
Machine F	Mechanical	06/21/2011 8:00	06/21/2011 00:00
Machine F	Hydraulic	07/11/2011 13:00	07/11/2011 00:00
Machine C	Coolant	07/13/2011 8:00	07/13/2011 00:00
Machine B	Coolant	07/15/2011 22:00	07/15/2011 00:00
Machine D	Coolant	07/27/2011 23:00	07/27/2011 00:00
Machine E	Mechanical	07/30/2011 8:00	07/30/2011 00:00
Machine C	Mechanical	10/10/2011 18:00	10/10/2011 00:00
Machine A	Mechanical	10/19/2011 11:00	10/19/2011 00:00
Machine D	Electrical	11/6/2011 16:00	11/6/2011 00:00
Machine E	Mechanical	11/8/2011 14:00	11/8/2011 00:00
Machine F	Mechanical	12/6/2011 8:00	12/6/2011 00:00
Machine F	Hydraulic	12/26/2011 13:00	12/26/2011 00:00
Machine C	Coolant	01/12/2012 8:00	01/12/2012 00:00
Machine B	Hydraulic	01/21/2012 8:00	01/21/2012 00:00
Machine B	Mechanical	02/4/2013 8:00	02/4/2013 00:00
Machine D	Electrical	02/13/2012 17:00	02/13/2012 00:00
Machine E	Electrical	02/17/2012 8:00	02/17/2012 00:00
Machine F	Mechanical	05/22/2012 20:00	05/22/2012 00:00
Machine F	Hydraulic	06/12/2012 8:00	06/12/2012 00:00
Machine A	Electrical	06/18/2012 22:00	06/18/2012 00:00
Machine B	Mechanical	07/30/2012 21:00	07/30/2012 00:00
Machine F	Mechanical	11/7/2012 10:00	11/7/2012 00:00
Machine F	Hydraulic	11/27/2012 15:00	11/27/2012 00:00
Machine A	Mechanical	05/3/2013 19:00	05/3/2013 00:00
Machine A	Hydraulic	05/28/2014 8:00	05/28/2014 00:00

According to the schedule in Table 11.22, no more than one job of preventive maintenance is to be performed per day, and the job of preventive maintenance is not to be performed every day in the proposed planning horizon. The proposed schedule for preventive actions avoids some system breakdowns, increases available production times, and reduces costs.

11.12 Stage 5: Data and Statistical Analysis Update

The proposed methodology objective was to develop a preventive maintenance schedule that addressed possible failures predicted from product defect times and types of defects. Statistical analysis was conducted to build models for machine failure and repair behavior modeling and multiple variable regression models. The schedule that was proposed in the previous section depended on data that was used for statistical modeling.

To have a complete and accurate preventive maintenance system, statistical models have to be updated continuously. This update incorporates monthly failure, repair, and defect times for use in statistical analysis to replace the current models. This can be done through a routine that updates the statistical models. The next chapter addresses this point in more detail.

11.13 Summary

A plan was developed in this chapter for preventive maintenance. This plan shows a time schedule in which a preventive action is to be conducted. The schedule addressed other functions in an organization such as production to achieve maintenance goals, and objectives of increasing production available time and reduce costs. A five-stage methodology was implemented in a case study of a local oil and gas company. The main production line of six machines in a series was chosen. There were four types of failures: mechanical, electrical, hydraulic, and coolant. Not all machines in the production line have the same type of failure. The first stage of the methodology was data preparation; data was obtained in a specific format in preparation for the next stage. The second stage was statistical analysis applied for distribution fitting analysis and multiple variable regression analysis. TBF and RT data were utilized to develop a statistical distribution for failure and repair times for each machine for each type of failure. Multiple variable regression was implemented to develop a regression model that predicts time of failure. Time of defect due to machine malfunction and dummy variables that represent the type of failure were the independent variables for the regression model, while the time of failure was the dependent variable. A regression model was developed for each machine. Only one machine in the regression model (machine F) showed that the type of failure is significant.

The models for the other machines predict the time of failure only according to the time of defect.

The third stage was system's simulation. A simulation model using WITNESS simulation software was developed. The model was called the as-is model, which describes the current production line. Machines were modeled in this instance to address different failure and repair behaviors through statistical models that were developed in the statistical analysis stage. A new simulation model called "defect time" was developed. This model addressed the generation of DTs for each machine to generate PFTs for a corresponding machine. According to the TBDs for each machine calculated from historical data, a code was used to generate DT once a part left a machine. The validated regression models were used to calculate the PFT for each machine using the generated DT. PFTs cover a planning period of four years. The regression models predict the failure time without predicting the type of failure, with the exception of machine F. Further analysis is needed to determine the probability of the predicted failure for each type of failure. The time between predicted failures and predecessor failures for the corresponding machine for each type of failure and the TBF statistical distributions were used to calculate the probability of failure for each type of failure. The type of failure that has the highest probability was considered for scheduling purposes. A list of PFTs and type of failure with highest probability of occurrence for each machine was established and used in the fourth stage of the methodology known as maintenance scheduling.

Preventive maintenance actions were scheduled to be performed on a third shift while production was not scheduled; each PFT was moved back to the time of the closest third shift. Performing preventive maintenance action on this shift prevents a failure while a machine is running. This preventive maintenance action also prevents future random failures to occur. Preventive actions were performed for each machine for a particular system (type of failure) within the time of the PFT, so it was not performed on each machine for each type of failure, as this would become unreasonable in terms of time and cost; preventing a failure increases the available time of the production line and reduces overall cost.

In conclusion, breakdowns of production systems generate high costs that reduce competitive capability. An integrated methodology for preventive maintenance was developed to increase preventive actions over corrective actions. A schedule was established for predicted failure times. Incorporating the developed methodology increases the time a production system is available and reduces costs. For the predicted model to stay synchronized with a system's behavior, a routine can be considered that automatically updates data used in the statistical analysis stage. Machine behaviors in terms of failure and repair can change as a machine deteriorates over time, so statistical distributions of TBF and RT should be updated. The regression model also needs to be updated, as it, too, may change with time. Updating depends on the complexity of the production line, but it is not recommended that updating be done, say, every month or in short periods of time because it will be tedious and ineffective task to do.

Author

Hazem J. Smadi is an assistant professor of industrial engineering at Jordan University of Science and Technology. His research interest is applied statistics, quality, reliability, and maintenance management.

Dr. Smadi has a BSc and MEng in industrial engineering from the University of Jordan. He earned his PhD from University of Houston. He is editorial assistant for the *International Journal of Rapid Manufacturing, and the International Journal of Collaborative Enterprise.*

References

Ahuja, I., and J. Khamba. 2008. Total productive maintenance: Literature review and directions. *International Journal of Quality & Reliability Management* 25(7): 709–56.

Davis, M., 2010. Contrast coding in multiple regression analysis: Strengths, Weaknesses, and Utility of popular coding structures. *journal of Data Science* 8(1): 61–73.

Garag, A., and S.G. Deshmukh. 2006. Maintenance management: literature review and directions. *Journal of Quality in Maintenance Engineering* 12(3): 205–38.

Mehdi, R., Nidhal, R., and C. Anis. 2010. Integrated maintenance and control policy based on quality control. *Computers and Industrial Engineering* 58(3): 443–51.

Panagiotidou, S., and G. Nenes. 2009. An economically designed, integrated quality and maintenance model using an adaptive Shewhart chart. *Reliability Engineering and System Safety* 94(3): 732–41.

Panagiotidou, S., and G. Tagaras. 2007. Optimal preventive maintenance for equipment with two quality states and general failure time distribution. *European Journal of Operational Research* 180(1): 329–53.

Simeu-Abazi, Z., and Z. Bouredji. 2006. Monitoring and predictive maintenance: Modeling and analyses of fault latency. *Computers in Industry* 57(6): 504–15.

Smadi, H. 2011. An Integrated Methodology for Preventive Maintenance Planning. Ph.D. diss., University of Houston, Houston.

Index